EU Regulation of GMOs

BIOTECHNOLOGY REGULATION SERIES

Series Editor: Han Somsen, *Tilburg Institute of Law, Technology and Society (TILT), Tilburg University and Centre for Environmental Law, University of Amsterdam, The Netherlands*

Biotechnology is a term that provokes a range of differing reactions and views, though most would agree on the necessity of ensuring its careful regulation. The rise and rapid evolution of modern biotechnology has generated serious and diverse regulatory challenges, many of which have significant implications for society as a whole.

This series is designed to comprise work, both collaborative and single-authored, that provides a critical insight into the challenges presented by biotechnology, and the range of regulatory techniques and solutions on offer. The issues confronted in the series will range from agricultural biotechnology, including GMOs, to human genetics, through to intellectual property aspects of biotechnology, such as patents and the TRIPS framework. The European and international dimensions of biotechnology regulation will be a constant reference point.

Whilst focusing principally on the legal framework of biotechnology regulation, the series will also draw from related disciplines such as environmental studies, politics and biology, and aims to inform policy as much as to comment on it.

Titles in the series include:

The Regulatory Challenge of Biotechnology
Human Genetics, Food and Patents
Edited by Han Somsen

EU Regulation of GMOs
Law and Decision Making for a New Technology
Maria Lee

EU Regulation of GMOs

Law and Decision Making for a New Technology

Maria Lee

Professor of Law, University College London, UK

BIOTECHNOLOGY REGULATION

Edward Elgar

Cheltenham, UK • Northampton, MA, USA

Published by
Edward Elgar Publishing Limited
The Lypiatts
15 Lansdown Road
Cheltenham
Glos GL50 2JA
UK

Edward Elgar Publishing, Inc.
William Pratt House
9 Dewey Court
Northampton
Massachusetts 01060
USA

A catalogue record for this book
is available from the British Library

Library of Congress Control Number: 2008932901

ISBN 978 1 84542 606 4 (cased)

Typeset by Cambrian Typesetters, Camberley, Surrey
Printed and bound in Great Britain by MPG Books Ltd, Bodmin, Cornwall

Contents

Abbreviations

ACRE	Advisory Committee on Releases to the Environment
AEBC	Agriculture and Environment Biotechnology Commission
BSE	bovine spongiform encephalopathy
Bt	*Bacillus thuringienesis*
CFI	Court of First Instance
DEFRA	Department for Environment, Food and Rural Affairs
DNA	deoxyribonucleic acid
EC	European Community
ECJ	European Court of Justice
EFSA	European Food Safety Authority
EPC	European Patent Convention
EPO	European Patent Office
EU	European Union
GATT	General Agreement on Tariffs and Trade
GM	genetically modified/genetic modification
GMO	genetically modified organism
JRC	Joint Research Centre
NAFTA	North American Free Trade Association
SPS	sanitary and phytosanitary
TBA	Technical Board of Appeal
TBT	technical barriers to trade
TRIPS	Agreement on Trade Related Intellectual Property Rights
v-CJD	variant Creutzfeldt-Jakob disease
WTO	World Trade Organization

Series editor's preface

Advances in biotechnology raise regulatory issues that both quantitatively and qualitatively differ from anything that regulators have had to deal with in the past. *The Regulatory Challenge of Biotechology*, the first book in this series, explored some of the most pressing of these challenges in agricultural and medical biotechnologies.

The title of this monograph, *EU Regulation of GMOs,* might suggest that this second volume in the series represents a highly specialised account of the regulation of genetically modified organisms in the European Union. Indeed, GMO regulation, especially within the foggy institutional context of the European Union, is spectacularly technical and complex, and hence to some extent invites such a specialised account.

However, one major argument that Maria Lee forcefully articulates in this volume, is precisely that stalemates in the regulation of GMOs in part are the inevitable result of exclusive reliance on different kinds of specialisation such as scientific risk assessment and patent law, and on European technocracy more generally. It is thefore particularly important that this book, apart from containing a strong argument in favour of public involvement at various stages of the regulatory process, amounts to the kind of accessible and holistic account of current EU policy as regards GMOs that is needed to facilitate such public participation.

Although this volume concerns EU law pertaining to GMOs, each chapter is doing justice to the reality that its proper understanding necessitates an appreciation of the institutional and political context from which those legal provisions arise. This is true in respect of risk regulation, labelling, coexistence, and the EU's regime on the patentability of GMOs. Finally, although not so long ago many EU lawyers could perhaps still afford to remain ignorant about WTO law, this certainly no longer applies to those who deal with GMOs. Chapter 6 therefore most appropriately is devoted to the WTO, which operates from a paradigm that is disctinctly different from that of the EU, which is likely to continue to remain a source of conflict for the foreseeable future.

In sum, there is every prospect that this book will de-mystify and expose the fascinating world of GMO regulation to a new audience and provide a new perspective for those already familiar with the subject. That this is much needed will be hardly open to debate after this important book.

Han Somsen
Tilburg Institute for Law, Technology and Society (TILT)

Acknowledgements

Colleagues and students at University College London, King's College London and elsewhere have given great support during the writing of this book. I am as ever especially indebted to Carolyn Abbot and Joanne Scott for the generosity with which they share their time and insights. The importance of Tanya Aplin's help with the rigours of intellectual property and Joanne Scott's patient guidance in WTO law cannot be overstated, and I am very grateful to Thijs Etty for his careful consideration of parts of this book. Luke Adams, Nep Elverd and Suzanne Mursell at Edward Elgar Publishing have all been fantastically helpful during the writing and production of this book. Special thanks to Steve Fownes for all his help.

Table of cases

EUROPEAN UNION

European Court of Justice

Court of First Instance

Commission Decisions

NATIONAL CASES

Canada

United Kingdom

United States of America

ORGANISATIONS

European Patent Organisation (EPO)

WTO: Panels and Appellate Body

Table of legislation

EUROPEAN UNION

Regulations

DIRECTIVES

Council Decisions

Commission Decisions

NATIONAL LEGISLATION

Canada

Germany

CONVENTIONS AND OTHER INTERNATIONAL INSTRUMENTS

1. Introduction

Biotechnology astonishes in its ingenuity and its potential. It confers the ability to change the characteristics of living organisms potentially without limit, transferring genetic information and traits across species. It grants the capacity to increase control over our surroundings, benefiting the environment, farmers and consumers, as well as the very poor in developing countries. The transformative potential of agricultural biotechnology, however, cuts both ways, raising profound questions about the type of world we are creating. Resistance is neither surprising nor unreasonable. As Sheila Jasanoff puts it, 'these far-reaching alterations in the nature and distribution of resources, and in the roles of science, industry and the state, could hardly occur without wrenching political upheavals'.[1] The regulation of genetically modified organisms (GMOs) reflects a real tension that pervades the management of all new technologies, between a desire to reap economic and social benefits and concern about unintended consequences.[2]

The history of technological change might suggest that, although resistance to change is common, it can be overcome, with change eventually normalised. History also reminds us, however, that there can be a dark side to progress, not least the closing down of other possible responses to needs or wants. To take an apposite example, even if we cannot readily identify alternative paths to social benefits, regret for the (perhaps unthought-of) path never taken has to be possible in the face of the huge environmental damage that goes along with the benefits brought by current farming practices.

---

1 Sheila Jasanoff, *Designs on Nature: Science and Democracy in Europe and the United States* (Princeton University Press, 2005), p. 4.

2 Andrew Barry, *Political Machines: Governing a Technological Society* (The Athlone Press, 2001), observes 'a political preoccupation with the problems technology poses, with the potential benefits it promises, and with the models of social and political order it seems to make available', p. 2; also Monroe E. Price, 'The Newness of New Technologies' (2001) *Cardozo Law Review* 1886.

THE EU'S MORATORIUM ON AUTHORISATION OF GMOs

GMOs properly hit EU politics in the late 1990s,[3] receiving a huge amount of media attention. The regulatory framework for GMOs, in place since 1990, was perceived to be profoundly inadequate and was targeted by a range of groups including environmental interest groups and those representing small farmers. The late 1990s saw high-profile and popular campaigns against GM food around the EU. So, for example, in both the UK and France protestors destroyed GM crops.[4] Not only were their actions widely reported, but the protestors took advantage of subsequent criminal trials to highlight their concerns about GMOs.[5] Dolly the sheep, the first cloned mammal, was introduced to the international media in Scotland in 1996, stimulating enormous interest in the safety and ethical implications, as well as the potential, of biotechnology. In 1997, the European Commission went ahead with the authorisation of a variety of GM maize in spite of angry objections from a number of Member States, a European Parliament resolution against authorisation, and the positive approval of only one Member State in Council.[6] This notorious case highlighted not only possible overreaching on the part of the Commission, but also the real disagreement on the content and appropriateness of risk assessment under the 1990 legislation. And then in 1998 Dr Arpad Pusztai announced on a television documentary that rats fed on GM potatoes suffered from stunted growth, suppressed immune systems and reduced body weight. There was a very public battle about the validity of his data, culminating in his suspension from employment. Although his claims never really

---

[3] Other than in quotations, I will for simplicity refer to the EU/European Union rather than the EC/European Community even when discussing what is strictly still 'EC law'. Note that, following the 2007 Lisbon Treaty, the Union will in any event take over the Community's legal capacity.

[4] See the discussion at the beginning of Chapter 2. This continues around the EU. Crops have more recently been destroyed in at least France, the Netherlands and Germany; see European Commission, *Second Report on the Experience of Member States with GMOs placed on the Market under Directive 2001/18/EC on the Deliberate Release into the Environment of Genetically Modified Organisms* COM (2007) 81 final.

[5] This fits into what Hilson calls 'legal opportunity'; see Christopher Hilson, 'New Social Movements: The Role of Legal Opportunity' (2002) 9 *Journal of European Public Policy* 238.

[6] Commission Decision 1997/98 [1997] OJ L 31/69. This decision was subject to litigation in Case C-6/99 *Greenpeace v Ministère de l'Agriculture et de la Pêche* [2000] ECR I-1651. See Tamara Hervey, 'Regulation of Genetically Modified Products in a Multi-Level System of Governance: Science or Citizens?' (2001) 10 *Review of European Community and International Environmental Law* 321.

recovered, this case created prominent media coverage of scientific uncertainty about the effects of GMOs.[7]

The legal response to this enormous interest in GMOs was spectacular and unplanned. The legislation that had required the authorisation of GMOs since 1990 fell apart. Between 1998 and 2004 no applications for authorisation of GMOs reached the end of the decision-making process, and a number of Member States introduced measures barring national market access to GMOs that had already been authorised. This was the famous *de facto* moratorium on the authorisation of GMOs,[8] a remarkable, probably unprecedented, breakdown in the EU legal framework. The moratorium was 'implemented' by the Commission's decision to stop pushing GMOs through the authorisation process. And this was most immediately prompted by declarations from 12 (of the then 15) Member States that they were opposed to further authorisations of GMOs.[9] It is too soon to go into detail on EU decision-making procedures (on which see especially Chapter 3) but, in short, these Member States had the majority they needed in Council to reject Commission proposals for the authorisation of specific GMOs.[10] Although the legality of the Council's position would have been at least questionable under the terms of the legislation, the Commission chose not to attempt to challenge, or indeed regularise, the moratorium. Instead, the Commission worked with the Member States and others to renegotiate the regulation that applied to GMOs, completely replacing and strengthening the EU's legislative framework. Only after the deadline for

---

[7] Dr Pusztai gave evidence for the purposes of House of Commons Select Committee on Science and Technology, Session 1998–99, 1st Report, *Scientific Advisory System: Genetically Modified Foods.*

[8] The nature of the 'moratorium' is actually under-analysed; see Sarah Lieberman and Tim Gray, 'The So-called "Moratorium" on the Licensing of New Genetically Modified (GM) Products by the European Union 1998–2004: A Study in Ambiguity' (2006) 15 *Environmental Politics* 592. Note in particular that GM food continued to be authorised under the 'substantial equivalence' procedure under Regulation 258/97 concerning Novel Foods and Novel Food Ingredients [1997] OJ L 43/1; see Chapter 3.

[9] Two declarations were made by different groups of Member States in the 2194th Council Meeting, 24/25 June 1999: *Declaration by the Danish, Greek, French, Italian and Luxembourg Delegations Concerning the Suspension of New GMO Authorisations*; *Declaration by the Austrian, Belgian, Finnish, German, Netherlands, Spanish and Swedish Delegations.* With slightly different emphases, both declarations state the intention of the Member States to block the authorisation of GMOs in Council pending amendment of the legislation. Ireland, Portugal and the UK did not join either declaration.

[10] Decision 1999/468 Laying Down the Procedures for the Exercise of Implementing Powers Conferred on the Commission [1999] OJ L 184/23, allowing Commission proposals on authorisation to be rejected by qualified majority voting, entered into force on 18 July following the declarations.

implementation of the new regulation did the Commission begin to step up its formal pressure on Member States.[11]

Whilst cause and effect between moratorium and pressures on the Commission is not straightforward, the Commission was at this time making a determined effort to bolster its own legitimacy in other areas. It was in no position to face down the Member States on such a politically sensitive topic. Most obviously, a European Parliament Committee Report, detailing numerous allegations of nepotism and financial fraud and mismanagement, led to the resignation of the entire Santer Commission in March 1999.[12] Not surprisingly, this received massive media attention around the EU, and amplified scepticism about the Commission, the EU institutions and even the ideals of EU integration. The new Prodi Commission was not formally appointed until September 1999. So the immediate response to the declarations leading to the moratorium came at a time when the EU was being administered by what was effectively a 'caretaker' Commission.[13]

And this came on top of the credibility crisis provoked by BSE. In March 1996 the British Government had announced a link between 'mad cow disease' (that is bovine spongiform encephalopathy (BSE)), eating beef and a new form of a fatal and distressing human brain disease, Creutzfeldt-Jakob disease (v-CJD).[14] The possibility of such a link had always been vehemently denied by the UK Government and the Commission. The relevant British scientific working party into BSE had in 1988 concluded that it was 'most unlikely' that there would be any human health implications. It had emphasised uncertainty, and also the serious implications of being wrong, but this was nevertheless for a number of years interpreted by the British Government

---

11 By the time of the mid-term review of the *Life Sciences Strategy* (European Commission, Staff Working Document, *Communication on the Mid-term Review of the Strategy on Life Sciences and Biotechnology* SEC (2007) 441), two cases were pending, p. 44. The Commission had earlier brought a partially successful action against France in respect of failure to implement the legislation, C-296/01 *Commission v France* [2003] ECR I-13909.

12 Committee of Independent Experts, *Allegations Regarding Fraud, Mismanagement and Nepotism in the European Commission* (1999). See Paul Craig, 'The Fall and Renewal of the Commission: Accountability, Contract and Administrative Organisation' (2000) 6 *European Law Journal* 98.

13 The interim Marin Commission.

14 The full story can be found in the Phillips Report, *Inquiry into the emergence and identification of Bovine Spongiform Encephalopathy (BSE) and variant Creutzfeldt-Jakob Disease (vCJD) and the action taken in response to it up to 20 March 1996* (House of Commons, 2000), available at http://www.bseinquiry.gov.uk/ (accessed December 2007). See Elizabeth Fisher, *Risk: Regulation and Administrative Constitutionalism* (Hart Publishing, 2007), Chapter 2, for a concise and interesting examination of the many complexities.

as a basis for a 'campaign of reassurance' on the safety of beef.[15] BSE affected many thousands of animals in the UK, and the mass slaughter that followed provided dramatic and widely distributed images of pyres of burning cattle. Beef sales plummeted, along with public confidence in regulators. The disease was most prevalent in the UK, but quickly became a European problem as other Member States sought to limit its effects on their own industry, restricting British beef imports, either independently or through EU action. In 1997 the European Parliament criticised in great detail the Commission's handling of BSE, accusing the Commission amongst other things of a 'policy of disinformation' in respect of both public opinion and relations between Community institutions.[16] BSE also threatened the fundamental assumptions of the internal market as Member States refused to comply with EC law.[17] This was the context in which the first significant imports of GM products (soya) began in 1996. So, at a time of significant public interest in the import of GMOs, the basic competence of the EU institutions in matters of risk was an issue of popular politics.[18]

The BSE debacle stimulated a massive rethink of risk regulation around the EU and in the EU itself. The politics of risk, especially around food, has been stirred up afresh by more localised scandals from time to time, for example dioxin in Belgian poultry (arguably partially responsible for the subsequent fall of the Belgian Government), salmonella in eggs, foot and mouth disease and avian flu. Food risks are highly sensitive, with cultural and historical aspects that exacerbate the potential for disagreement around the EU. We should expect GM food to be no different, compounding the anxiety about the technology more generally, and spilling over into non-food GMOs such as cotton and flowers. The period of the moratorium saw a major upheaval in the

---

15 Phillips Report, above n. 14.

16 Medina Ortega Report, Temporary Committee of Inquiry into BSE, *Report on Alleged Contraventions or Maladministration in the Implementation of Community Law in Relation to BSE* (European Parliament, 1997). See Graham R. Chambers, 'The BSE Crisis and the European Parliament' in Christian Joerges and Ellen Vos (eds), *EU Committees: Social Regulation, Law and Politics* (Hart Publishing, 1999); Kieran St Clair Bradley, 'Institutional Aspects of Comitology: Scenes from the Cutting Room Floor' in Joerges and Vos, ibid. The Parliament Report was also heavily critical of the UK Government.

17 See Case C-241/01 *National Farmers' Union v Secretariat General du Gouvernement* [2002] ECR I-907; Case C-1/00 *Commission v France* [2001] ECR I-9989; Case C-393/01 *France v Commission* [2003] ECR I-5405; Case C-180/96 *UK v Commission* [1998] ECR I-3903.

18 The BSE crisis is explicitly linked to the moratorium on GMOs by Gregory Shaffer and Mark Pollack, 'Regulating Between National Fears and Global Disciplines: Agricultural Biotechnology in the EU' (2004) *Jean Monnet Working Paper* 10/04.

approach of the EU to food safety. A new General Food Regulation was introduced in 2002.[19] It sets out general principles of food law, and establishes the European Food Safety Authority (EFSA). Whilst agencies at national level usually 'borrow' the political legitimacy of the national political institutions, here the Commission seems to be hoping to gain some legitimacy from the exercise of expertise by the EFSA. According to the General Food Regulation, EFSA is 'an independent scientific point of reference in risk assessment'.[20] The Regulation uses EFSA to provide scientific excellence, and to separate this science from political (including national in this EU context) and industry influence. These are direct responses to some of the identified failures in the management of BSE.[21] The rethink of food risks, and its considerable centralisation through EFSA, is directly relevant to the perceived need to renegotiate the regulatory framework applying to GMOs. The Commission would probably be wrong to see public rejection of GMOs as being solely about food, although this is the primary focus at the moment.[22] The concerns, especially about uncertainty and the purposes or distribution of benefits of the technology, apply more broadly. But, nevertheless, food does end up in a very particular institutional framework that is especially sensitive to both the *political* risks and the *internal market* risks of getting things wrong. By asserting its role in the response to risk, the EU makes clear the potentially profound implications of the internal market. Different national approaches to risk pose major challenges to the internal market, as exemplified by the experience with BSE, and indeed again with GMOs. In addition, however, regulating risk is seen as a way to respond visibly to the needs of the European peoples, to re-engage with citizens. The institutions have become self-aware in their search for legitimacy, seeing risk regulation as, perhaps paradoxically, both a way to 'reconnect' with the European publics and a way to reduce threats to the internal market.

In the period of the moratorium, the EU's legitimacy problems were extensive and high profile, ranging from the 'democratic deficit' (which can imply concern about the very existence of the EU as well as its decision-making mechanisms) to questions about basic competence and questions of honesty and trustworthiness. One of the central responses to the EU's legitimacy

---

[19] Regulation 178/2002 Laying Down the General Principles and Requirements of Food Law, Establishing the European Food Safety Authority and Laying Down Procedures in Matters of Food Safety [2002] OJ L 31/1.

[20] Regulation 178/2002, above n. 19, Recital 34.

[21] The scientific committees dealing with BSE had been dominated by British experts, and heavily influenced by the British political agenda (in turn too closely involved with promotion rather than regulation of the industry). See above n. 16.

[22] See for example European Commission, above n. 11.

dilemmas has been an effort to move towards improved openness and public involvement in the 'life' of the EU.[23] There are many possible meanings for 'public participation', and many possible purposes, from improving democracy to encouraging regulatory compliance. In this book, public participation in decision making is not particularly examined in its own right. Instead, it is linked with the vast range of issues raised by GMOs, and with the need to explore values that go beyond those covered by risk assessment.

A turn to participation is now almost an instinctive response to concerns about legitimacy, concerns that, as we have seen, arise frequently and with no little drama in the EU. The Commission's 'European governance' project typifies the turn to participation.[24] The governance project was instigated in 2000, a self-conscious response to, bluntly put, the unpopularity both of the EU and of the Commission more specifically. The five 'principles of good governance' established in this process were 'openness, participation, accountability, effectiveness and coherence'.[25] An important and explicit part of Commission reflection at this period was a reassessment of the position of scientific and technical expertise in decision making.[26] A self-consciousness about 'scientific governance' permeates policy, and commitment to openness, especially with regard to uncertainty and disagreement, is a conventional element of the governance of expertise in the EU. The regulatory bodies seem to understand the importance of public participation, the importance of values as well as safety, to decision making. But there is real ambiguity on this, generally and in the regulation of GMOs, given the continued commitment to scientific decision making.

---

[23] Discussed in detail in Maria Lee, *EU Environmental Law: Challenges, Change and Decision-Making* (Hart Publishing, 2005), Chapter 5.

[24] See European Commission, *White Paper on European Governance* COM (2001) 428 final, and of the numerous other reports particularly European Commission, *Communication on the Collection and Use of Expertise by the Commission: Principles and Guidelines* COM (2002) 713 final. Note also the parallel project on Commission reform, European Commission, *Reforming the Commission* COM (2000) 200 final. Note that the Almaty Declaration (UNECE, 2005) expands the Aarhus Convention's (*Convention on Access to Information, Public Participation in Decision-making and Access to Justice in Environmental Matters*) principles on public participation to GMOs.

[25] Commission (2001), above n. 24, p. 10.

[26] As part of the governance project, above n. 24, European Commission, *Communication on the Collection and Use of Expertise by the Commission: Principles and Guidelines 'Improving The Knowledge Base For Better Policies'* COM (2002) 713 final. Also for example European Commission, *Science and Society Action Plan* (2002); European Commission, *Communication on the Precautionary Principle* COM (2000) 1 final.

Efforts to persuade rather than enforce were obviously attractive when the Member States declared their opposition to further authorisations of GMOs in 1999. Moreover, given the rethinking of risk and governance at this time, the adequacy of the legislation was suspect even to a techno-enthusiastic Europhile. Public anger about GMOs seemed to take everyone, not just the Commission, by surprise, as indeed does the continued persistence of concern. There was a sense that formal action would be counterproductive. Even the affected commercial entities hesitated,[27] and in any event food retailers and processors began to respond to consumer rejection of GM technology, in some cases undertaking that their products did not contain GM material. Most strikingly perhaps, whilst the Commission was under considerable pressure from its trade partners (especially the US) even before the moratorium, formal World Trade Organization (WTO) dispute resolution was not sought until 2003. This is not an enormous delay relative to other claims, although dissatisfaction with EU regulation predated the moratorium, but the delay did mean that the new regulatory structure was in place before litigation. We might assume that the hestitation was in part down to fear of further consumer backlash against the technology, and a hope that consumers would learn to accept GMOs. But, in addition, it was perhaps recognised that such a high profile and difficult dispute had the potential to disrupt already controversial WTO bodies. There had been massive protests during the WTO's ministerial conference in Seattle in November 1999 (not long after the Member State declarations on GMOs). These protests were so intense that the conference's opening ceremony was cancelled, a state of emergency declared and a curfew imposed.[28] The protests included a wide range of groups and interests, and were the most dramatic demonstration that the trade elite should henceforth expect public scrutiny of the impact of trade on other social objectives. Seattle became the pivotal moment in a much broader 'anti-globalisation' movement, which included environmental, development and consumer protection perspectives. GMOs had vast potential as a symbol for a particular negative perception of globalisation. The vulnerability of the international trading system's popular legitimacy and authority in the longer term emphasised for the EU at least the need to address the public concerns that had provoked the moratorium.

---

27 Although in 2001 Monsanto challenged Italy's safeguard measures (Case C-236/01 *Monsanto Agricoltura Italia SpA v Presidenza del Consiglio dei Ministri* [2003] ECR I-8105). In May 2007 (Case T-139/07 *Pioneer Hi-Bred v Commission* [2007] OJ C 155/28) Pioneer Hi-Bred challenged the Commission's failure to submit its application in respect of insect-resistant GM maize 1507 to committee. The industry also brought civil actions against the protestors digging up trial planting.

28 See Oren Perez, *Ecological Sensitivity and Global Legal Pluralism: Rethinking the Trade and Environment Conflict* (Hart Publishing, 2004), pp. 1–7.

But if all of this makes caution about GMOs seem the only possible response, we need to remember the very significant pressures that urge speedy commercialisation. 'Knowledge' is thought generally to be a key source of wealth in post-industrial society, and that must have enormous implications for government (EU or national) policy towards science and technological development. The *Life Sciences Strategy*, drafted by the Commission and 'welcomed' by the Council,[29] presents biotechnology as 'the next wave of the knowledge-based economy',[30] and a key objective of the EU is, famously, to be 'the most competitive and dynamic, knowledge-based economy in the world.'[31] The *Life Sciences Strategy* addresses 'white', 'red' and 'green' (industrial, medical and agricultural[32]) biotechnology policy together, which emphasises the size of the potential economic benefits associated with biotechnology.[33] A wide range of industrial sectors is identified as 'based on' biotechnology, which in turn is deemed essential for economic prosperity.[34] Indeed, the economic potential of biotechnology pervades the *Life Sciences Strategy*, and is if anything increasingly emphasised in the ongoing annual reviews of the Strategy: the overriding objective seems to be 'to improve the situation for European biotechnology'.[35]

---

29 *Council Conclusions and Roadmap of 26 November 2002 for a Strategy on Life Sciences and Biotechnology* [2003] OJ C 39/9. The Communication was considered by industry and competitiveness councils, whilst the environment council considers applications for authorisation of individual GMOs.

30 European Commission, *Life Sciences and Biotechnology – A Strategy for Europe* COM (2002) 27 final (the *Life Sciences Strategy)*, p. 7. The *Life Sciences Strategy* has been subject to annual 'progress reports'; see COM (2003) 96 final, COM (2004) 250 final, COM (2005) 286; see also European Commission, *Communication on the Mid-term Review of the Strategy on Life Sciences and Biotechnology* COM (2007) 175 final.

31 Lisbon European Council, 2000. Links are explicitly made between the 'Lisbon Strategy' and the 'Life Sciences' in both strategies. The Lisbon strategy was updated in 2005 to focus on 'stronger, lasting growth and more and better jobs' (European Commission, *Working Together for Growth and Jobs: A New Start for the Lisbon Strategy* COM (2005) 24 final).

32 'Blue' biotechnology, the marine and acquatic applications of biotechnology, is occasionally discussed.

33 European Commission, above n. 30, constantly emphasises the economic value of the biotechnology. The UK's Better Regulation Commission, *Risk, Responsibility and Regulation – Whose Risk is it Anyway?* (2006) cites the figure (from 'one source', unidentified in the report) of £2 billion as the negative impact on the UK economy of the 'controversy surrounding GM crops', p. 18.

34 Les Levidow and Clair Marris, 'Science and Governance in Europe: Lessons from the Case of Agricultural Biotechnology' (2001) 28 *Science and Public Policy* 345.

35 COM (2004) 250 final, p. 2. 'there is a strong need to continue promoting the development of life sciences and biotechnology in the EU', *Mid-term Review*, above n. 30, section 6.

The whole meaning and transformative nature of the 'knowledge economy' is contested, especially here because the role of agricultural biotechnology in this brand new economy is as much about potential as current performance. Whilst the contribution of biotechnology to economic development tends to be presented in the official policy as objective fact, a necessity to which we must adapt, these conclusions are of course profoundly value laden, closing out alternative development paths. But when biotechnology is presented as an economic revolution (and by 2007 the Commission is talking of the 'bio-economy'[36]) the consequences of being left behind start to look disastrous. This economic focus means that corporate and government priorities increasingly coalesce around questions of 'wealth creation'.[37]

There are other possible social benefits to agricultural biotechnology, beyond the economic, as hinted in the first paragraph of this chapter, and as explored a little further in Chapter 2. But whilst feeding the poor and (especially) sustainability are increasingly referred to in EU policy, it is the sense that an economic opportunity is passing us by that dominates. Like the transformative potential of GMOs, the economic potential cuts both ways. It is a huge promise, but the profit motive also creates mistrust and provides no reason to tolerate uncertainty.

2001 saw the introduction of the first major plank of the new structure of regulation, with the Deliberate Release Directive,[38] applying to all GMOs for release into the environment or placing on the market. By 2003 the other major pieces of legislation were also in place – the Food and Feed Regulation,[39] applying special rules to GM food and feed, and the Traceability and Labelling Regulation,[40] filling out the rules on labelling and traceability for all GMOs.

---

36 European Commission, above n. 11. Indeed the Commission has even set up a network on the knowledge-based bio-economy – KBBE-NET, ibid.

37 Joseph Murphy and Les Levidow, *Governing the Transatlantic Conflict Over Agricultural Biotechnology: Contending Coalitions, Trade and Standard Setting* (Routledge, 2006). The 'Competitiveness in Biotechnology Advisory Group', consisting of representatives 'from all the various industry segments and from companies at every stage of company development together with entrepreneurial academics' made recommendations to the Commission, above n. 30, p. 7. Industry (and 'entrepreneurial academics') have a great deal to contribute on enhancing competitiveness and wealth creation, but there is an obvious risk that they (and these particular elements of policy) could dominate.

38 Directive 2001/18 on the Deliberate Release into the Environment of Genetically Modified Organisms and Repealing Council Directive 90/220/EEC [2001] OJ L 106/1.

39 Regulation 1829/2003 on Genetically Modified Food and Feed [2003] OJ L 268/1.

40 Regulation 1830/2003 concerning the Traceability and Labelling of Genetically Modified Organisms and the Traceability of Food and Feed Products

In theory the 'moratorium' on authorisations was brought to an end by the authorisation of Bt11 sweetcorn in May 2004.[41] It is not at all obvious that the decisions taken so far really demonstrate the existence of an effective and predictable regulatory system. Under the new legislation, Member States can either accept or reject the Commission's proposal on an application for authorisation of a GMO by qualified majority voting. In every case until the end of 2007, the Member States have been deadlocked, unable to find a qualified majority in either direction.[42] That means that the Commission takes the final decision on authorisation. This way of ending the moratorium is no less acrimonious or ambiguous than its inception.

THE NEW TECHNOLOGY

Technology powerfully affects our relationships with each other, and with our environment, redistributing risks and benefits and potentially changing the ways we think about the world. One of the serious difficulties in the regulation of GMOs is in the extreme polarisation of views on what is at stake with this technology. Some think that this is an unprecedented change in human relationships with our environment, others that it is a simple next step in our constant efforts to control the environment around us. Even the 'newness' of GMOs is contentious.

Manipulation of plant and animal genes for human ends has been going on for millennia, albeit until recently in ignorance of the existence of the gene. Traditional breeding involves 'crossing' plants or animals in the search for a preferred trait, such as increased productivity. Similarly, ancient applications of technology to food production include the use of enzymes in the preparation of food and drink, most obviously the micro-organisms in yeast in brewing beer and baking bread. Indeed the phrase 'agricultural biotechnology' sometimes implies a continuum with modern applications of technology. In this book, 'agricultural biotechnology' is used in a more popular sense, specifically to refer to modern biotechnological techniques. In any event, I start from the position that we should simply be thinking much harder about modern applications of biotechnology than we have to about, for example, the use of

Produced from Genetically Modified Organisms and amending Directive 2001/18/EC [2003] OJ L 268/24.

41 Commission Decision 2004/657/EC authorising the placing on the market of sweetcorn from genetically modified maize line Bt11 as a novel food or novel food ingredient under Regulation (EC) No 258/97 of the European Parliament and of the Council.

42 The Decisions can be found at http://ec.europa.eu/food/dyna/gm_register/index_en.cfm (accessed December 2007).

yeast. This book is not about all agricultural biotechnology applications, but specifically about the regulation of GMOs.[43] The manipulation of living organisms at the level of the gene rests on the 19th century discovery of the inheritability of the characteristics of living organisms by Mendel. Watson and Crick described the double helix in 1953,[44] and from that point the gradual discovery of the properties of 'genes' led to the 1973 turning point of insertion of toad DNA into a bacterial cell. In modern biotechnology, barriers of sexual compatibility, which had always meant, for example, that a bacterium could not be 'crossed' with corn, no longer apply. Modern biotechnology can isolate the DNA fragments responsible for a preferred trait, manipulate that gene in the laboratory, and insert it into potentially any other organism.

Social and political tensions around new technological development are not unique to genetic modification. One of the questions about the experience of GMOs is whether more general lessons can be learned for the future. Lessons are, and should be, sought in past experience, even if one of those lessons is that it is not sensible to extrapolate directly from one technology to another, for example, from nuclear power or chemicals policy to agricultural biotechnology.[45] Differences between technologies rest not only on the hazards and benefits of a new technology, but also on the social processes that they engage. The acceptability of a technology depends on a number of factors, including the familiarity and personal usefulness of the technology (for example, information and communications technologies as opposed to biotechnology), the distribution of knowledge (so whether we can get decent technical advice from our neighbour or only from more distant experts) as well as risks and benefits, and the open or monopolistic structure of the industry controlling the technology.[46] It should not be assumed that a new technology can be assessed using the tools that proved effective for the last technology. Familiar and elaborate

---

43 See Eleni Zika et al., *Consequences, Opportunities and Challenges of Modern Biotechnology for Europe* (JRC, European Commission, 2007) for a review of the many different technologies available.

44 James Watson and Francis Crick, 'A Structure for Deoxyribose Nucleic Acid' (1953) 171 *Nature* 737. DNA (deoxyribonucleic acid) is a molecule that contains the information that controls the synthesis of enzymes and other proteins, and they then provide the basic metabolic processes of all cells. A gene is a DNA sequence, and the total set of genes of an organism is the genome. This is organised into chromosomes in the nucleus of the cell.

45 Gregory N. Mandel, 'History Lessons for a General Theory of Law and Technology' (2007) 8 *Minnesota Journal of Law, Science and Technology* 551, discusses the efforts of 19th century courts to accommodate the telegraph (a new technology) within existing legal categories.

46 Robin Grove-White, Phil Macnaghten and Brian Wynne, *Wising Up: The Public and New Technologies* (Centre for the Study of Environmental Change, Lancaster University, 2000).

tools of risk assessment failed to capture the issues relating to genetic modification. And the approach taken (or missed) in respect of biotechnology will not necessarily capture what is distinctive next time. But, nevertheless, there seems to be a particularly self-conscious reflection on the experience of biotechnology in discussion of nanotechnology, perceived as the 'next big thing', both in its positive potential and, apparently, in its potential to provoke public concern.[47] Anticipation of widespread public concern in this case contrasts sharply with what seems to have been a complete failure of regulators and the industry to anticipate the intensity of the public response to the new technology of genetic modification.

In the case of nanotechnology, there may still be opportunities for engagement before institutional and economic commitments are entrenched. This is one big lesson from the agricultural biotechnology experience – the conditions that might make the technology socially acceptable were considered far too late. Engagement on GMOs was through protest, 'objection' rather than a positive consideration of the development of social life. The UK Government's approach to nanotechnology seems to accept the fundamental point that early engagement is needed. Government agrees it should initiate 'adequately funded public dialogue around the development of nanotechnologies', and that 'properly targeted and sufficiently resourced public dialogue will be crucial in securing a future for nanotechnologies':

> The Government's aim for public dialogue around nanotechnologies is to elicit and understand people's aspirations and concerns around the development of these technologies. Through the dialogue process, scientists and the public can jointly explore existing and potential opportunities, and policy-makers will want to hear about, and then respond to, public concerns related to ethical, social, health, safety and environmental issues.[48]

This is potentially very positive. We will see throughout this book that there is a basic mainstream acceptance of decades of work and mountains of evidence from the social sciences on public responses to new technologies and risk.

---

[47] See for example Royal Society and Royal Academy of Engineering, *Nanoscience and Nanotechnologies: Opportunities and Uncertainties* (2004). See also some of the documents on the EU's nanotechnology website (http://cordis.europa.eu/nanotechnology/home.html, accessed December 2007), where the use of biotechnology experience is more implicit, although sometimes raised directly, for example *The Future of Nanotechnology: We Need to Talk* (2006). More generally, the Council sees research in biotechnology as 'a model for integrating activities addressing ethical and social aspects from the earliest possible stage', above n. 29, para. 6.

[48] Department of Trade and Industry, *Response to the Royal Society and Royal Academy of Engineering Report: 'Nanoscience and Nanotechnologies: Opportunities and Uncertainties'* (DTI, 2005), para. 80.

Different approaches by experts and the public are not necessarily about ignorance or irrationality, but might include judgments on the broader context of the developments. But as we will also see throughout this book, there are significant hangovers from the deeply embedded understanding of expert approaches as real and rational and public approaches as emotional and irrational.[49] And, even at a purely practical level, putting the new understanding into action is enormously difficult. So for nanotech we need to think about how we might realistically perform this upstream engagement, when research is in the private sector, and its objectives and trajectories are unclear. Public involvement generally is more difficult to achieve when decisions are distant. The perceived relevance to one's own life increases as the decision gets closer in time and in space. Regardless of the amount of debate on nanotechnology 'upstream', there may well be a new constituency when a decision is made on using nanotechnology in the hospital, factory or field down the road.

And even in the report quoted above, the UK Government takes a somewhat instrumental view towards public participation, such that public involvement makes an already determined trajectory for a technology more acceptable. Government also, even at this early stage, whilst acknowledging social and ethical concerns, concentrates on more tractable (although still difficult) questions of risk to the environment and human health. Whilst the relevance of a broad range of values is accepted, the familiar questions of safety and risk return to the centre of things. We will see this throughout our discussion of GMOs. In part this might be government paying cynical lip service to the importance of democracy. In part, it might just be that this really is very difficult. Brian Wynne argues that whilst the 'political fact' of public opinion is accepted as real and indeed legitimate, the reality, the 'intellectual substance' of what lies behind it, is not. Public views are not 'recognized to be what they are, which is public judgments of the quality of existing knowledge, and of the exaggerated claims made for it by scientists and the policy bodies they advise'.[50] Official emphasis on science silences these concerns about uncertainty, and by presenting the 'objectivity' of decisions leaves unspoken the economic and scientific values behind decisions.[51] It is not possible to simply tack the correct understanding of public views onto existing ways of doing things.

---

49 See for example Brian Wynne, 'Creating Public Alienation: Expert Cultures of Risk and Ethics on GMOs' (2001) 10 *Science as Culture* 445.

50 Wynne, above n. 49, p. 457.

51 See Joanna Goven, 'Processes of Inclusion, Cultures of Calculation, Structures of Power: Scientific Citizenship and the Royal Commission on Genetic Modification' (2006) 31 *Science, Technology and Human Values* 565.

Although the potentially disruptive effects of new technologies can be exaggerated, genetic modification does pose new and difficult problems for lawyers. This book addresses legal responses to the dilemma of agricultural biotechnology, ranging from the authorisation process, through intellectual property to international trade law. The law in this area addresses modes of decision making as much as substantive standards, and modes of decision making are forced to grapple with the promise and threats of GM. Whilst this book examines the legal framework of decision making, however, it should be clear that the social acceptability, the meaning, of new technologies cannot be reduced to a series of discrete and relatively simple decisions.

THE STRUCTURE OF THE BOOK

A range of legal disciplines are inevitably involved in the regulation of GMOs. There is a danger that each legal discipline will pass the buck, claiming that the most difficult legal questions are simply not relevant in that area. One of the central, if often implicit, concerns of this book is the need for dialogue between different legal specialisms. So, for example, it is increasingly accepted that GMOs will have distributional effects, and that the way in which a technology distributes costs and benefits will affect its acceptability. But the full magnitude of the dislocation faced by non-GM farmers, and its almost complete neglect by the framework for authorisation, is brought sharply into focus by the potential scope of patent infringement. Neither patent law nor the set of rules on authorisation obviously has the tools to consider this dislocation comfortably. This might turn attention to the role of civil liability in redistributing the costs and responsibilities of GMOs, but here we find an opaque set of rules no better suited to addressing these questions. Only a more holistic approach to the regulatory framework can begin to identify and examine legal responses to GMOs. In particular, the aim of this book is to examine the capacity of these different legal disciplines to hear and address the range of issues raised by these new technologies. Laws and regulations a-plenty are in place, and institutions are adapting and shifting to deal with the newness of GMOs. But GMOs provoke a very broad range of concerns, and it is debatable to what extent law is able to hear, let alone answer, the full range of social questions.

This book attempts to examine what is at stake in the regulation of GMOs. Chapter 2 outlines the main pressures faced in the regulation of agricultural biotechnology, the range of hopes and concerns raised by the technology. The safety of GMOs is not straightforward, but poses relatively familiar and manageable problems for regulators. Safety questions also imply, however, real issues of ignorance and uncertainty, which are less tractable. This chapter also explores a related and less precise category of political, ethical and socioeconomic concerns about GMOs. These include distributional issues, food security, and biocentric questions about the integrity of nature. The range of concerns and the underlying uncertainty about the impacts of the technology demand that we engage with the role and purpose of GMOs. We will see as we move through the chapters that there is a constant struggle to find space in the regulatory process for the full expression of the substantive questions legitimately raised in respect of GMOs. But the jurisdictions examined here have at least recognised in principle the value-based nature of debate on new technologies. A far-reaching and innovative 'GM dialogue' in the UK provides a case study in Chapter 2 of a government trying to come to terms with the breadth of issues raised by agricultural biotechnology, through a combination of expert and public deliberation.[52] This GM dialogue also shows another thread running through this book. Even if the breadth of the debate is recognised in principle, risk or at least expertise always seems to be the most comfortable basis for the decision. The hegemony of risk, discussed also in this chapter, is very difficult to shake off.

Chapter 3 discusses in detail the authorisation process for GMOs in the EU. EU legislation requires the industry to seek authorisation of each GMO on a case-by-case basis, and this chapter traces the efforts of legislators to respond to diverse and conflicting pressures in this area. As discussed above, the legislation on GMOs fell apart quite spectacularly in the late 1990s, with the introduction of a moratorium on the authorisation of GMOs. The new set of legislation attempts to respond to the inadequacies of the earlier framework. As suggested in the discussion of Chapter 2 above, one of the lines running through this book is the difficulty of addressing in the regulatory framework the full breadth of questions provoked by GMOs. This is reflected in a tension between the role of science (and sometimes also other forms of expertise) and the role of politics or public opinion in decision making on GMOs, and compounded in the EU authorisation regime by the sensitivity of the allocation of authority between national and central decision makers. This question of the battle for authority in a system of multi-level governance is another of this book's themes. Chapter 3 explores the detail of the legislation along these lines of debate, and in short argues for greater national autonomy on GMOs, allowing national decision making on the basis not just of risk to the environment or health, but also on the basis of 'other legitimate factors'.

The 'regulation' of agricultural biotechnology does not begin and end with the procedure for the authorisation of a product or process, and Chapters 4 and

---

[52] Generally in this book, whilst of course other jurisidictions are of great interest, the UK is used when a national example is required.

5 move into the broader scope of regulation, looking at the legal environment that applies when authorised GMOs are cultivated or marketed in the EU. The regulation of GMOs in the EU is underpinned by a rhetoric of consumer choice, given legal substance by obligations to label GMOs. It seems to be assumed that any public concerns about GMOs not addressed in the authorisation process can be picked up in the market place. And, because we have mandatory labelling, the market does provide a space for some expression of the full range of objections to GMOs. It is, however, unfortunate that the pattern of the legislation concentrates so much on the largely individualised and shallow forum of the market. The role of consumers in GMO regulation crucially assumes the availability of meaningful choice, and so some distinctiveness between GM, conventional and organic farming. But the pervasiveness of genetic material means that, once cultivation of GM crops is widespread in the EU, GM material is likely to be present in non-GM crops.

Rules to ensure the 'coexistence' of different forms of agriculture are in principle a matter for the Member States, but Chapter 4 explores the very tight (and, it is argued, largely misconceived) constraints within which the national authorities are operating. The potentially serious impact of GM farming on other forms of agriculture, for example if the presence of GM material affects the status (organic or 'GM-free') or performance of non-GM food or crops, raises questions of civil liability for adverse effects of GM farming. It is currently far from certain in what circumstances GM farmers or the biotechnology industry will face liability for the adverse effects of their industry. Nor is liability for other possible negative impacts (on human health or the environment) either entirely clear or likely to be extensive. The allocation of costs and responsibilities if things go wrong should have been a central part of the very hard fought, slow and complex regulatory settlement on GMOs. Liability rules are always politically difficult, but they are especially fraught in respect of GMOs for two main reasons. First, like other regulation, liability prefers to deal with calculable risk. But *uncertainty* about effects pervades the regulation of GMOs. And, secondly, much of the detail of liability schemes provides an implicit protection for innovative or regulated products on the assumption that they can be deemed to be in the public interest. In fact, consensus around the public interest in this particular technological development is precisely what has broken down.

Intellectual property in agricultural biotechnology is explored in Chapter 5, as a further final element of the legal environment in which GMOs are grown and sold. The proprietary nature of the technology and the dominance of large corporate ownership highlight questions about the distributional impact of GMOs, globally and locally. The assertion of control over its products by the biotechnology industry in this context contrasts very sharply with its complete denial of responsibility for untoward effects. It is also interesting to note that the patenting system is so far the most fruitful area of regulation for consideration of the ethical implications of agricultural biotechnology. But again, whilst the language is in place, it is extraordinarily difficult to follow through.

In short, the current framework for 'living with GMOs' leaves some issues entirely unaddressed, and responds poorly to others. Much of the policy discussion seems to assume that widespread GM agriculture is an objective and unavoidable fact to which other forms of agriculture must adapt, rather than a choice.

The EU regime on GMOs was developed and is applied in the shadow of WTO rules on trade liberalisation. This has led to an acrimonious, longstanding and ongoing disagreement over the propriety of EU regulation of GMOs. North American and other farmers who had adopted GM crops widely in the 1990s found themselves unable to sell their products in the EU, and the biotechnology industry found an enormous market drifting out of reach. There is also a sense that there was much to play for, with the US and EU each trying to influence the choice of regulatory regime elsewhere in the world. Chapter 6 analyses some of the main challenges for the EU in justifying its GMO regulation before the WTO, including discussion of the 2006 Panel decision on the moratorium. The WTO rules do provide considerable space for the justification of otherwise doubtful measures by reference to a broad range of public values, and space for most of the concerns expressed about GMOs could be found somewhere. But some issues are heard more clearly than others. There are many incentives on WTO members to frame the reasons for their legislation modestly, primarily on the basis of risk to human health and environmental protection. So, as before, notwithstanding an openness in principle, the actual basis for a decision is in fact rather narrow. Even these social objectives must be pursued by WTO members within very tight constraints. As at other levels of decision making, the degree of autonomy for the members of the WTO to respond to the demands of their citizens is a challenging question. The development of a reasonably predictable and testable approach to decision making, without discounting democratic pressures, is an important task and suggests, as at EU level, the need to work through greater flexibility for members.

The regulation of GMOs is about far more than science, risk and safety. GMOs provoke a wide range of substantive concerns which merit serious consideration. This implies in turn that the public as well as experts have a role in regulation. But an overwhelmingly scientific (or technical, or even legalistic) focus can silence other voices. The danger of impoverishing debate is real.

2. GMOs in the EU: the scope of the debate

INTRODUCTION

The law applying to GMOs is deeply embedded in its social context. The breadth as well as the depth of concerns about agricultural biotechnology, and the resulting ambiguity about what is at stake, create interesting challenges for law and for those charged with reaching decisions in particular cases. This chapter explores in some detail the scope of the debate on GMOs, beginning with an outline of some of the main arguments in favour of the technology as well as the main concerns, before turning to a discussion of the hegemony of risk as a basis for decisions. The innovative and extensive process of review and public participation carried out in the UK during the regulatory hiatus created by the EU moratorium provides a case study.[1]

A number of GM 'events' begin to provide a sense of the scope and the depth of concern about GMOs. In 1998 and 1999 a series of high-profile 'direct actions' took place against GMOs in the UK, as individuals and groups destroyed trial fields of GM crops, in front of both cameras and the police. Lord Melchett (the then executive director of Greenpeace) used the media attention to raise what he saw as the 'very serious threat to the environment' posed by GMOs, as well as Greenpeace's concern about both the institutions controlling the technology and the institutions controlling the decision making: 'the chemical industry think they have some sort of private right to plant this stuff and cause this pollution to the environment and this threat . . . the public have a right to protect the environment'.[2]

In France, similar protests have had an even higher profile. As recently as December 2007, José Bové and a number of other protestors used the opportunity of their sentencing for destruction of GM crops to announce a hunger

[1] See Chapters 1 and 3 for discussion of the moratorium. The UK did not support the EU moratorium, but entered into a voluntary moratorium by agreement with industry.

[2] BBC News website 29 July 1999, http://news.bbc.ca.uk/1/hi/uk/406191.stm (accessed December 2007).

strike in the cause of a one-year ban on GMOs in France.[3] As in the UK, the threats and uncertainties of GMOs and their control are highlighted in these protests. In addition, the French protests, spearheaded by the Confédération Paysanne, positioned GMOs firmly as a threat to the maintenance of local and culturally distinctive food, and firmly in the sights of the anti-globalisation movement.[4]

Also in 1998, the dispute over scientific evidence in the 'Pusztai' affair made headlines in the UK for weeks. Dr Arpad Pusztai announced on a *World in Action* television documentary that rats fed on GM potatoes suffered from stunted growth, suppressed immune systems and reduced body weight. He was suspended from his employment, to general media excitement, and his data were disputed (and have never been fully rehabilitated). The very public conflict brought to centre stage a range of credible concerns about the uncertainty of the impact of GMOs, and highlighted the sometimes controversial nature of scientific claims. Doubts about the safety of GM food have been raised persistently since. In 2007, for example, *Le Monde* published details of a claim that rats fed with Monsanto's MON863 corn suffered negative effects on the liver and kidneys. Highlighting the imbalance of knowledge and information between regulator and industry, these conclusions were based on a reinterpretation of the company's own data, which Greenpeace had obtained through a German court order.

A further prominent 'event' in the politics of genetic modification arose in famine threatened southern Africa in 2002.[5] Food aid provided by the US contained GMOs that had not been authorised by the recipients. A number of countries accepted the aid only on condition that the seeds were first milled. The US refused to mill it, claiming it was too expensive, but South Africa

---

3 A four-month prison term was commuted to a fine, Reuters, 10 December 2007, http://uk.reuters.com/article/environmentNews/idUKL1011368420071210 (accessed December 2007).

4 See http://www.confederationpaysanne.fr/ (accessed December 2007). José Bové has been imprisoned before for direct actions – and quite dramatically, with helicopters and a dawn raid to avoid his supporters. Crops have also been destroyed in at least the Netherlands and Germany; see European Commission, *Second Report on the Experience of Member States with GMOs placed on the Market under Directive 2001/18/EC on the Deliberate Release into the Environment of Genetically Modified Organisms* COM (2007) 81 final.

5 Institute for the Study of International Migration, *Genetically Modified Food in the Southern Africa Food Crisis of 2002–2003* (Georgetown University School of Foreign Service, 2004). This report illustrates that the situation is far from straightforward. See also Jennifer Clapp, 'The Political Economy of Food Aid in an Era of Agricultural Biotechnology' (2005) 11 *Global Governance* 467; Noah Zerbe, 'Feeding the Famine? American Food Aid and the GMO Debate in Southern Africa' (2004) 29 *Food Policy* 593.

stepped in. Whilst the countries demanding milling continued to state concern about the health impacts of GM maize consumption, the milling of the whole-kernel maize meant that no GM seeds could be planted, reducing the risk of crossing with local varieties. Zambia, however, would not accept the GM food aid at all, arguing that the population's high consumption of maize (which is a staple in southern Africa, so that comparison with US consumption is not necessarily helpful) could lead to unanticipated health impacts, and also that lack of resources would make it difficult to address any problems.[6] The US accused the EU, which was in the middle of its moratorium, of contributing to hunger – the countries of southern Africa had a strong interest in remaining GM-free in order to maintain exports to the EU. Others retaliated with accusations that the US's food aid policies were economically motivated, and that the US was trying to force GM agriculture into Africa. The mud-slinging, and at the very highest diplomatic levels, was extraordinary.[7] And it placed the African food shortage firmly in the context of EU/US disagreement over agricultural biotechnology. For current purposes, the moral outrage around this case highlights the distributional questions inherent in the export of agricultural biotechnology. This case also emphasises the impact of globalisation on local control of decision making, with the future of farming and food production decided very far from where the impact of those decisions will be felt.

Together, these episodes dramatise concerns about human health and the environment, which are underlined and massively complicated by profound scientific uncertainty. The authority and legitimacy of decision making are disputed, and concerns about the distribution of the costs and benefits of GMOs are compounded by the effects of globalisation. We are looking not just at questions of perception or fear, things that might be turned around by persuasion, education or reassurance. These issues are complex, substantive, nuanced and deep-rooted. Later chapters will examine the extent to which the legal frameworks applying to GMOs allow or enable the full range of concerns to resonate in regulatory decisions, particularly at the EU and World Trade Organization (WTO) levels. Finding a regulatory home for the deliberation of the full range of perspectives is far from easy, and a constant theme of this book.

---

6 Addressing errors such as 'Starlink' in the US has cost millions of dollars; see below, p. 28.

7 For a taste, see Institute for the Study of International Migration, above n. 5. Joseph Murphy and Les Levidow, *Governing the Transatlantic Conflict over Agricultural Biotechnology: Contending Coalitions, Trade and Standard Setting* (Routledge, 2006), quote letters published in the *Wall Street Journal* from European Commissioners and from US Trade Representative Robert Zoellick, pp. 156–7.

THE GM DEBATE

Disagreement about the appropriate role of GMOs in agriculture and in food production is intense and sometimes bad-tempered. Those supporting GMOs paint those resisting as irresponsibly disregarding the greater public interest, and opponents characterise proponents similarly. There are many and powerful reasons for staying alive to the potential of GMOs, and relatively few of those resisting the technology are against it always and forever. The purpose of the technology, its governance and the control of its impact are all important factors.

Of the extensive range of concerns about GMOs, not all are readily accepted as legitimate, or as suitable subjects for regulation. The very identity of the problem that law and regulation has to solve is up for grabs.[8] Concerns about safety (broadly environmental and health issues), whilst challenging, can be incorporated into regulation in a way familiar from many other technological developments. More problematic is a less precise category of 'other' (political, ethical and socioeconomic) issues, which also requires the attention of regulators. The division is not as neat as that, and there are significant dangers in carving out the technical and the 'other'. Most obviously, the 'categories' overlap: environmental and health impacts raise ethical and political questions around the distribution of risks and benefits, for example. This is particularly so for long-term and unpredictable risks, and the inevitable uncertainty about the effects of GMOs underscores the burden of justification on those introducing the technology. Moreover, if presented as objective and inevitable fact, the scientific element of a decision can hide from view the value and ethical judgments pervading the decision itself, so that it looks as if only the protestors are making such judgments.[9] We might also note here that 'values' (or public or social values, also public goods[10] or collective goods) are perhaps overused terms, which will be used much more in this book. These terms often describe 'public' perspectives on the world (rather than private or individual interests), which are developed in the context of deliberation around

---

8 Julia Black, 'Regulation as Facilitation: Negotiating the Genetic Revolution' (1998) 61 *Modern Law Review* 621.

9 Brian Wynne, 'Creating Public Alienation: Expert Cultures of Risk and Ethics on GMOs' (2001) 10 *Science as Culture* 445; Joanna Goven, 'Processes of Inclusion, Cultures of Calculation, Structures of Power: Scientific Citizenship and the Royal Commission on Genetic Modification' (2006) 31 *Science, Technology and Human Values* 565.

10 No term of art is implied here, but to economists public goods are goods that are non-excludable (withholding benefits from one means withholding from others) and non-rival (consumption by one does not reduce their availability).

the public interest, rather than preformed in a self interested context.[11] In this book, the intention is rather more simply to capture the full range of issues raised by GMOs. Environmental and human health protection are important values as is the pursuit of economic prosperity, but discussion of values goes beyond questions of risk, embracing also the 'other' issues discussed here. And again, it must not be thought that only protestors have 'values'. The purposes of GMOs, controlling the environment or growing the economy, are value based and should be equally subject to scrutiny.

Why Push Ahead? The Danger of (Pre)caution

Most generally, the precedent set by abandoning or partially abandoning a technology would challenge deep-rooted and successful ideas about progress through science. And the economic promise of GMOs is substantial, not only in its own right, but also as part of the great post-industrial hope of the 'knowledge economy', as discussed in Chapter 1.

But equally importantly, the claimed and potential benefits of GM technology seem to be limited only by the imagination, and include advantages in respect of all the major risks identified by those who oppose agricultural biotechnology, that is, environmental, health and 'other' socioeconomic, political and ethical benefits. Biotechnology beyond the farm has myriad actual and potential applications,[12] but here the emphasis is on farming with GMOs. Policy makers are increasingly emphasising benefits as they come to terms with the continuing reluctance of the EU public to accept biotechnology in agriculture,[13] and it is probably true that demonstrable benefits would influence the willingness to accept risks and uncertainties, and other difficulties in this area. If the social benefits do not materialise, however (and so far, on the whole they have not) problems of credibility will multiply.

The hoped-for environmental benefits of GM technology could address many of the problems associated with contemporary farming. The development of plants that do not need fertilisers (perhaps because they fix nitrogen from the air), for example, would avoid the pollution of water sources by

---

11 Mark Sagoff, *The Economy of the Earth: Philosophy, Law and the Environment* (Cambridge University Press, 1988), provides the classic exposition for environmental lawyers.

12 See Eleni Zika et al., *Consequences, Opportunities and Challenges of Modern Biotechnology for Europe* (JRC, European Commission, 2007) for discussion.

13 See for example European Commission, *Life Sciences and Biotechnology – A Strategy for Europe* COM (2002) 27 final (the *Life Sciences Strategy*), and European Commission, *Communication on the Mid-term Review of the Strategy on Life Sciences and Biotechnology* COM (2007) 175 final. For a positive, but reasonably sober, review of what is going on, see Zika et al., above n. 12.

fertilisers. The negative impacts of irrigation (soil salination, water shortages) could be avoided by GM drought-resistant plants. The modification of animals or their feed could lead to a reduction of the amount or noxiousness of effluent. Very closely targeted pest resistance could end the loss of biodiversity associated with chemical pesticides. And improved yields could mean that less land needs to be given over to agriculture, even as improved adaptability means that more marginal land could be farmed. The pragmatic appeal of these possibilities is immense, even if they are largely in the future, and even if individual 'technical fixes' fall short of addressing the fundamental nature of ecological harm.

Biotechnology could also respond to past environmental harms. One of the seminal 'patenting life' cases, *Diamond v Chakrabarty,* involved a bacterium engineered to break down crude oil.[14] Genetic modification of crops identified as useful biofuels could increase availability or efficiency and reduce reliance on fossil fuels. The ability to engineer plants to thrive in harsh environments (drought-resistant, salt-resistant wheat with strong short stalks to withstand storms) is on some accounts urgent as we face human-induced climate change. EU policy makers talk a lot about the potential for GM technology to buy our way out of climate change.[15] The ambiguity of progress is unavoidable – scientific and material advance has caused unintended increases in the level of greenhouse gases in the atmosphere, leading to potentially horrific climate change, but a new and equally transformative technology promises the possibility of life more or less as we know it. Again, the pragmatic appeal of the technical fix is enormous.

The modification of agricultural products to address the health problems of the wealthy world, for example by reducing the cholesterol content of dairy products or providing appetising child-friendly flavours for healthy vegetables, is another promise of agricultural biotechnology. GM non-food crops could also provide novel ways to manufacture pharmaceuticals, by synthesising drugs or hormones more cheaply or efficiently than is otherwise possible, or allowing drugs to be eaten rather than injected. This is perhaps a more compelling use of new technology than chocolate flavoured broccoli. More controversially, genetic engineering of animals could increase the supply of organs for transplant through xeno-transplantation, by reducing the chances that animal organs will be rejected by the human body. Non-food crops could contribute to cleaner or more efficient industrial manufacturing, as fields replace factories, producing plentiful and clean raw materials for industrial processes, such as biodegradable plastics. A turn to these alternative uses of

---

14 *Diamond v Chakrabarty* 447 US 303, 100 S Ct 2204, discussed in Chapter 5.
15 Above n. 13.

agricultural land diversifies the rural economy. Even in much of the developed world, agriculture is a troubled industry and diversification or improved production could just tip the balance for marginal employment.

The promise of a 21st-century 'green revolution', providing the poor with a means to feed themselves and to escape hunger and poverty, is difficult to ignore. Modifications of staple crops have the potential to provide improved yields (for example, through pest or disease resistance) or enhanced nutrition (by changing the nutritional content of staple food) and to make more marginal land available for growing food. The risk that such innovations will be lost is the darkest spectre faced by those who resist the development of GMOs. As the Nuffield Council on Bioethics puts it, '"More food for the hungry", unlike "tomatoes with longer shelf-life", is a strong ethical counterweight to set against the concerns of the opponents of GM crops'.[16]

However, the notorious food surpluses of the developed world over several decades raise questions about the possible contribution of GMOs to ending hunger: poverty, inequality, lack of access to land are as likely to be a cause of hunger as lack of food – lack of demand rather than lack of supply.[17] Many would be willing to accept that hunger is in part, in some times and places, a problem of underproduction. The current commercial reality of agricultural biotechnology, however, is that it focuses on two GM traits (herbicide tolerance and insect resistance) in a small number of commodity crops (mainly corn, soy, oilseed rape/canola and cotton). Almost by definition, these benefit farms with high chemical inputs, so more commercial, probably larger, farms. Most biotechnology research is not about feeding the poor or saving the environment, but is driven by the market and by economic objectives. This commercial imperative, a focus on 'wealth creation', can apply to government-funded as well as private sector research.[18] Even the most commonly cited example of the promise of GM, 'golden rice', suggests that the challenges are

---

16 Nuffield Council on Bioethics, *Genetically Modified Crops: the Ethical and Social Issues* (1999); see www.nuffieldbioethics.org/ (accessed December 2007).

17 See Amartya Sen, for example, *Poverty and Famines: An Essay on Entitlement and Deprivation* (Oxford University Press, 1981). For a provocative discussion of the potential of increased supply, see Mark Sagoff, 'Biotechnology and Agriculture: The Common Wisdom and its Critics' (2001) 9 *Indiana Journal of Global Legal Studies* 13.

18 Even in respect of public funding, the 'wealth creation' objective of scientific research can dominate, and can favour particular industrial sectors. In respect of the UK, see Agriculture and Environment Biotechnology Commission (AEBC), *What Shapes the Research Agenda? In Agricultural Biotechnology* (AEBC, 2005), especially para. 3.5.1. See also Tony Blair, 'Britain's Path to the Future – Lit by the Brilliant Light of Science', 3 November 2006, The King's Centre, Oxford, in which the then Prime Minister emphasised the commercial advantage that can be provided by science.

not simply technological, and even go beyond mistrust of the motives of those controlling the technology. Vitamin A deficiency contributes to childhood diseases and blindness in many children with diets heavily dependent on rice. Golden rice is modified (by the insertion of genes from the daffodil and bacteria) to contain higher levels of beta carotene, which converts into vitamin A. Critics claim however that the distribution of vitamin A pills could reduce vitamin A deficiency at much lower cost, or, better still, that diets should be improved through mixed farming, including local green vegetables that are less available when agriculture concentrates on commodity crops.[19]

The commercial imperative underlying research and development undermines the ethical argument from the proponents of agricultural biotechnology. Non-profit organisations (rich universities, or charities and developing nation governments[20]) could be pressured to fill this gap, and China and India seem also to be special cases, technologically advanced states facing many of the challenges of a developing country. And although the promise of agricultural biotechnology for the very poor remains largely unmet, and the continued absence of progress is likely only to increase cynicism, benefits for the very poor remain a real possibility.[21] Indeed, the possibility of feeding the world through GM is probably no less likely or imminent than mass redistribution of resources within and towards the developing world. It demands some humility. The same applies to the environmental potential of GMOs, currently somewhat hypothetical, but a possibility that merits consideration.

Safety Concerns

One dominant set of concerns about GMOs focuses on safety, primarily of the environment and of human and animal health.[22] The possible negative envi-

---

[19] See the discussion in Thomas Bernauer, *Genes, Trade, and Regulation: The Seeds of Conflict in Food Biotechnology* (Princeton University Press, 2003); Bill McKibben, *Enough: Genetic Engineering and the End of Human Nature* (Bloomsbury Publishing, 2003), pp. 141–4. The angry response provoked by Dick Taverne, 'The Real GM Food Scandal' *Prospect* November 2007, berating those objecting to GMOs for preventing the commercialisation of innovations like golden rice, can be followed at http://www.prospect-magazine.co.uk/list.php?category=171&issue=620 (accessed December 2007).

[20] Sagoff, above n. 17.

[21] The Nuffield Council reviews existing modifications that could benefit developing countries, above n. 16, para. 4.29. For a more pessimistic view, see Friends of the Earth International, *Who Benefits From GM Crops?* (FoE, 2006).

[22] An interesting place to look for a detailed examination of the main concerns is the UK Science Review: GM Science Panel, *GM Science Review: First Report: An Open Review of the Science Relevant to GM Crops and Food Based on the Interests and Concerns of the Public* (2003).

ronmental impacts of GMOs are disputed, but potentially serious. Some of the familiar environmental problems associated with conventional intensive farming could be intensified by GM technology. The two commercially dominant modifications involve herbicide-resistant and pesticidal crops.[23] So for example, adding a gene from the *Bacillus thuringienesis* (Bt) soil bacterium, which is toxic to certain pests, to a crop plant such as maize or cotton creates a plant with pesticidal properties. Other crops are modified to resist the application of certain herbicides (Monsanto's 'Roundup Ready' crops are best known), so that weed-killer can be applied at any stage of the growing cycle without killing the crop. Herbicide or pesticide tolerance is a serious problem in conventional farming, coming about as susceptible individuals die off and pests or weeds that survive breed resistance into their population. It requires farmers to turn to other pesticides and herbicides, leading to ever escalating chemical supplements, with ecological and economic impacts. It is a phenomenon that could be intensified by the widespread cultivation of crops modified to allow for constant exposure to pesticides or herbicides. Bt resistance is particularly worrying because Bt is one of the few pesticides allowed in extreme cases to organic farmers, so that widespread resistance could deprive organic farmers of a last-resort pesticide. Further, and again as with conventional pesticides, there are indications that crops modified to kill particular pests may be toxic to certain non-target species, most notoriously the discovery in laboratory tests that pollen from Bt corn can harm the larvae of the iconic (in the US) monarch butterfly.[24]

There are many other environmental concerns about GMOs. 'Clean' fields, free of weeds and pests, would have an obvious impact on farmland biodiversity: what are birds to eat in this promised land? There is also a downside to the potential to make ever more marginal land accessible to agriculture, since this could increase agricultural land take even further from current levels.[25] The spectre of 'superweeds' has also received a great deal of attention. The

---

23 See the worrying discussion of Bt resistance measures in the US, Rebecca Bratspies, 'The Illusion of Care: Regulation, Uncertainty, and Genetically Modified Food Crops' (2002) 10 *New York University Environmental Law Journal* 297.

24 John E. Losey, Linda S. Rayor and Maureen E. Carter, 'Transgenic Pollen Harms Monarch Larvae' (1999) 399 *Nature* 214. For a discussion of the initial report and subsequent criticisms of it, see Sheila Jasanoff, *Designs on Nature: Science and Democracy in Europe and the United States* (Princeton University Press, 2005) pp. 108–9. Whilst the effect of GM crops in the field is vigorously debated, it is accepted that pest-resistant crops can be harmful to non-target wildlife in the laboratory: GM Science Panel, *GM Science Review: First Report: An Open Review of the Science Relevant to GM Crops and Food Based on the Interests and Concerns of the Public*, para. 6.3.1, available at http://www.gmsciencedebate.org.uk (accessed December 2007).

25 This should be placed alongside the promise that increased yields reduce the amount of land needed for agriculture, given growing populations.

concern is that a modified gene could enter the natural environment, by cross-pollination with wild relatives, or as a volunteer, that is a crop that self-seeds and persists outside the field environment. This new plant could be difficult to control, if the genetic modification is for herbicide tolerance, or it could out-compete native wild plants, if modified for traits such as tolerance of pests, frost and/or drought.

The existence and gravity of risks to human health from the cultivation or consumption of GMOs are, like the environmental risks, disputed, but potentially serious.[26] There is concern that a modified gene could enter cells in the human gut, leading to new and unthought-of diseases. Less dramatically, unanticipated nutritional differences between GM and non-GM food could lead to widespread but initially faintly perceptible effects on public health. And allergic reactions to a particular GM food (or to the pollen of GM crops) are certainly possible: the modification of soybeans by a gene from the Brazil nut to make them more nutritious (introducing methionine, a nutrient that soybeans lack) also transferred a major food allergen from nuts to soybeans.[27] And finally, the difference between non-food and food maize, for example, is not apparent to the human senses. If GM crops not suitable for human consumption enter the human food chain, there is a possibility of harm, and again there is a notorious example. Because of concerns about allergenicity, 'StarLink' maize was approved in the US only for use as animal feed, and explicitly not for human consumption. When StarLink maize was discovered in the human food chain, there was an outcry, mass withdrawals of products from the market and enormous associated costs, although it is claimed that there were no impacts on human health.[28] If 'biopharming', that is, modifying crops to grow the raw materials for industrial or pharmaceutical processes, becomes commercially viable and widespread, the risks from unwanted material in the human food chain are intensified.

---

26 It is notoriously difficult to identify the long-term impact of food; see GM Science Panel, above n. 22, Chapter 5.

27 Stephen Nottingham, *Eat Your Genes: How Genetically Modified Food is Entering Our Diet* (Zed Books, 1998), p. 92.

28 See Rebecca Bratspies, 'Myths of Voluntary Compliance: Lessons from the StarLink Corn Fiasco' (2003) 27 *William and Mary Environmental Law and Policy Review* 591. StarLink corn was also found in food aid to Bolivia, but in this case the efforts to recall it were less dramatic; see Clapp, above n. 5. The EU legislation takes steps to avert any potential StarLink crisis, by requiring all authorised GMOs to meet the requirements for both food and feed: 'experience has shown that authorisation should not be granted for a single use, when a product is likely to be used for food and feed purposes; therefore such products should only be authorised when fulfilling authorisation criteria for both food and feed', Regulation 1829/2003/EC on Genetically Modified Food and Feed [2003] OJ L 268/1, Recital 10.

One final concrete controversy has arisen over the use of an antibiotic resistance gene as a 'marker' in GMOs.[29] The genetic modification of a plant can be carried out by inserting the desired gene into plant cells, which are then regenerated into a plant. A marker is used to identify successfully transformed cells, allowing breeders to preserve desirable strains. It is very common to use an antibiotic resistance gene as a marker. But concern has intensified in recent years about overuse of antibiotics and the development by bacteria of resistance to antibiotics, which would make the treatment of disease by that antibiotic ineffective. That general concern has become a part of opposition to GMOs. In particular, if bacteria living in the gut could pick up the resistance there could be a serious impact on human health.

Many of these safety issues involve side effects of industrial agriculture and food production that are vaguely familiar, and not even entirely unexpected. They are very difficult to regulate, but do not necessarily challenge our ordinary way of doing business. The most intractable issue around new technology is the nature of scientific uncertainty. Everything we know about GMOs lies in the shadow of ignorance, the prospect of harm that we have not even thought of. By definition we cannot test for these impacts, so the difficulty goes far beyond 'data gaps', problems that can be resolved through more or better science. This is as much a social as a scientific problem,[30] but our regulatory systems, as explored further below and in the next chapter, generally discount ignorance. The very good reason for this is that proof of safety is simply impossible, and our world would look rather poorer had we declined innovations because of that inability to prove safety. However, it is odd to resolve not even to contemplate the possibility of harms we have not thought of. We might hold those uncertainties in the account against the benefits of a new technology, particularly if those benefits are also uncertain. Understating or normalising uncertainty, as our regulatory systems often do, means also assuming the objectivity and inevitability of choices made on a scientific basis. This in turn diverts attention from the value assumptions and commitments of the regulatory framework.

A further limitation to our scientific knowledge arises from the release of GMOs into unpredictable and variable real-world social and ecological systems. Human behaviour is not predictable, and precautions will not always be taken. We have seen this in all areas of regulation, including GMOs. And knowledge of the operation of ecological systems is at best incomplete and fragmented. The cumulative impact of genetic modification is extraordinarily

---

29 This is explicitly addressed in the EU regulation; see Chapter 3, p. 74.

30 Brian Wynne, 'Uncertainty and Environmental Learning: Reconceiving Science and Policy in the Preventive Paradigm' (1992) 2 *Global Environmental Change* 111.

complex, and every time a different modification is introduced, additional complexity is introduced. 'Gene stacking' could result from unsought crosses of GM varieties with each other or with conventional crops or wild relatives, with a complex range of 'advantages' (or disadvantages) for the resulting plant. The real-world conditions cannot be replicated in the laboratory, and so impacts are not readily 'tested' in advance.[31] This 'indeterminacy' of effects is a further and profound form of uncertainty that scientific risk assessment can barely begin to tackle.[32]

Brian Wynne consistently emphasises the importance of these different and profound types of uncertainty. He identifies a rejection of the confidence expressed in institutional predictions as a central part of the public response to GMOs.[33] Whilst 'risk' along identified parameters (pesticide resistance, allergic reactions, and so on) is complex and likely to be disputed for many years, it is a familiar debate, and fits reasonably readily into familiar political and legal frameworks. The same cannot be said of more profound types of scientific uncertainty.

Other Concerns

As in respect of many new technologies (although in this case perhaps more visibly so), there are persistent concerns about agricultural biotechnology that are not directly related to the environmental/health risks discussed above. Public debate on GMOs in the UK revealed that 'the public do not view GM as purely a scientific, or environmental, or economic, or political or ethical issue. All of these aspects are important to them'.[34]

GMOs have become one of the focal points of the anti-globalisation movement. The challenges of new technologies, which by definition have the potential to bring with them unknown risks and unthought-of benefits, are compounded by the speed with which new technologies and new risks cross borders. Whilst the anti-globalisation movement has been derided as lacking clear objectives, it is hardly surprising that the very rapid increase in move-

---

31 The limitations of laboratory testing tend to be raised by environmental groups challenging the confidence of regulators and industry. We should note, though, industry reliance on this argument in response to the allegations about the impact of Bt corn on the larvae of the monarch butterfly (above n. 24); see Jasanoff, above n. 24, p. 109. This is picked up by the WTO *EC–Biotech* Panel; see Chapter 6, p. 218.

32 Again following Wynne, above n. 30.

33 Wynne, above n. 9.

34 *GM Nation? The Findings of a Public Debate* (AEBC, 2003), para. 42, emphasis omitted. The GM Nation website (www.gmnation.org.uk) is no longer available, but the report can be found on the AEBC website, http://www2.aebc.gov.uk/aebc/reports/gm_nation_report_final.pdf (accessed December 2007).

ment of goods, capital and technology around the world provokes political upheaval. Increased intensity of global interactions raises serious questions about the survival of democratic, culturally individual decision making, and about the distribution of power and resources. These questions are highly visible in the case of GMOs, especially with increasingly globalised scrutiny of domestic regulation through institutions such as the EU and the World Trade Organization, discussed in Chapters 3 and 6.

Agricultural biotechnology is thought likely also to enhance corporate control over agriculture, in particular because the capital resources required for research and development seem to point towards large corporate dominance.[35] Historically, agricultural research and development has been the domain of the public sector, but from the late 1980s it moved into the private sector. This was consistent with an increased reliance on the private sector and the market to provide public goods at that time, and is also linked with the development of intellectual property protection, discussed in Chapter 5. Biotechnological innovation also allows greater control by a single supplier over different elements of agriculture, as, for example, the same company supplies the herbicide-resistant seed and the herbicide.

Intensive conventional agriculture has already taken us quite far from local relationships between consumers and producers, but further intensifying transnational corporate influence shifts control of farming further towards more remote, profit-driven, corporate agendas. Impacts are likely to be felt far from where decisions are made. This is particularly difficult for developing countries, but European farmers may also find themselves squeezed between powerful suppliers and a concentrated retail sector. Agricultural biotechnology raises questions about how the risks and benefits of new technology will be distributed between industry, farmers and consumers. A related point is that, because they are a capital-intensive form of agriculture, GMOs are thought likely to encourage agricultural intensification and an increase in monocultures. Single crop species are usually more vulnerable to pests and disease than mixed farming, raising food security issues as well as environmental problems. And as ever, the poor are most vulnerable to shocks. It is also possible that, rather than providing benefits for the poor, agricultural biotechnology

---

35 On the question of corporate control as a central feature of concern, see Emma Hughes, 'Dissolving the Nation: Self-Deception and Symbolic Inversion in the GM Debate' (2007) 16 *Environmental Politics* 318, p. 325 and following. Joyce Tait and Joanna Chataway, 'The Governance of Corporations, Technological Change and Risk: Examining Industrial Perspectives on the Development of Genetically Modified Crops' (2007) 25 *Environment and Planning C* 21, discuss the long lead times and multinational corporation dominance of the 'innovation trajectory for the sector as a whole', p. 24.

will contribute to the undercutting of developing world agriculture by developed world agriculture, for example as tropical produce is modified to grow in cold climates.

A striking illustration of the ways in which farmers may be tied into an increasingly capital intensive system is in the development of 'terminator' gene technology (one form of 'genetic use restriction technology', or GURT).[36] A terminator gene renders harvested seed sterile, preventing the saving of seed for sowing in subsequent years, and similar technology could prevent breeding from GM animals. GURTs are familiar from the development of hybrids in the 20th century, which provide improved yields but are not suitable for breeding.[37] Farmers are prepared to purchase hybrid seeds every year for the benefits of the technology. By contrast, the terminator gene *in itself* provides no benefit for farmers. Although a terminator gene is likely to be marketable only if another genetic modification provides benefits worth the annual purchase, freedom of choice could be eroded by limited competition between seed companies, as well as by national government policies or international aid. Moreover, the gene itself benefits only the seller. The implications of terminator technology for poor or subsistence farmers in the developing world are potentially huge. Purchasing seed is a significant investment for poor farmers, and doing so every season increases the riskiness of farming. If marginal benefits of the GM crop turn out not to match the costs, poor farmers face devastating problems. Whilst the risks are less extreme, terminator technology could pose difficulties for farming in the developed world, especially small-scale farming. Terminator technology also has drawbacks for plant breeding, potentially further concentrating this industry, as well as ruling out small-scale improvements to seed by individual farmers and farming communities. And the spread of sterile seed into the natural environment or into a neighbouring crop could be catastrophic.

This particular technological advance clearly advantages the powerful, and binds farmers into a financial obligation to buy seed every year. On the other hand, terminator technology is said to have the potential to prevent gene spreading, averting a potentially significant negative environmental impact of GMOs. Environmental interest groups, however, reject this as a role for terminator technology, arguing that the technology is not sufficiently reliable to

---

[36] See the discussion in Graham Dutfield, 'Should We Regulate Technology through the Patent System? The Case of Terminator Technology' in Han Somsen (ed.), *The Regulatory Challenge of Biotechnology: Human Genetics, Food and Patents* (Edward Elgar, 2007).

[37] They are not necessarily sterile. Although they do not 'breed true', they can be used for breeding, and often are, perhaps being regenerated with newly purchased seeds every few years rather than every season.

prevent negative environmental impacts, even if it can work well enough to prevent seed-saving being economic. The potential negative impact of terminator technology is recognised in two decisions under the Convention on Biological Diversity, which call for more research on the environmental and socioeconomic impact of GURTs.[38]

The distributional impacts of GMOs cut both ways. Regulation in the EU has potentially profound impacts beyond its borders, and the US certainly alleges that the EU's position on GMOs contributed to hunger in southern Africa in 2002. Even more than environmental and health impacts, the socioeconomic impacts of GMOs are difficult to predict, escape experimental calculation, and are probably impossible to recall once they have happened. The intention here is not to choose between different views but to emphasise the importance of the debate. Urbanisation and changes to the rural economy have complex roots, and historically, technological change in agriculture has obviously not been straightforwardly bad. But technological change in farming does seem to encourage urbanisation and the flight of small farmers, as capital replaces land and labour as the central factor in agriculture. The detailed effects of the last 'green revolution' remain controversial, but it at least benefited richer farmers earlier than others, and some locations still lag behind. And, unlike the last green revolution, this one is dominated by the private sector and subject to property rights. Given that the focus here is on EU regulation, the complex and disputed impacts of the genetic revolution on developing countries can only be touched on. It might be observed, however, that questions of global distribution should form a legitimate element of the regulatory debate in the EU.

Many of the fears about the socioeconomic impact of GM crops are tied to an underlying concern that, through unwanted contamination of non-GM seeds (by cross-pollination, spilling of seed, mixing in processing), or because the structure of the seed supply industry makes sourcing of non-GM seeds increasingly difficult, virtually all farmers will practise GM agriculture and be subject to the conditions of the owners of the technology. The Soil Association argues that, once commercial growing of GM crops was widespread, it very quickly became difficult to purchase non-GM maize, soy and oilseed rape on the United States market, and even more so in respect of rape in Canada.[39] The concern that opportunities to avoid GM agriculture, dominated by a powerful corporate agenda, may shrink quite rapidly is related to another overarching factor of socioeconomic concern, which also focuses on distributional values.

---

[38] Conference of the Parties V, Decision V/5 (2000), and Conference of the Parties VIII, Decision VIII/23 (2006).

[39] Gundula Meziani and Hugh Warwick, *Seeds of Doubt* (Soil Association, 2002), Chapter 7.

Ulrich Beck highlighted some time ago the ways in which technology distributes risk and benefit, and in the picture he draws of the 'risk society' the distribution of bads (rather than goods) becomes the focus of political life.[40] At the moment, it seems as if the risks of this particular technology are likely to go to the more vulnerable (including developed-world publics and future generations, as well as poor farmers), and the benefits to the economically powerful producers. In these circumstances, the grand claims of those pushing the technology look weak, and the debate continues to be shaped by the question of what the technology is really *for*.

The pervasiveness of GMOs raises more specific questions, including consumer choice and difficulties for farmers who wish to market their produce as being free of GMOs, particularly organic farmers.[41] If organic farming is simply an economic choice, allowing profit to be made from romantic ideas of connection with nature, history is littered with industries that fail when faced with a new competitor. Even if this is all that is at stake, however, there should be the opportunity to debate whether the particular form of connection with nature provided by organic farming is worth the price of limitations on GMOs. But the issues are rather different in any event. First of all, the new industry does not simply out-compete the existing industry. By the pervasiveness of its product, GM farming makes it difficult for the existing industry (organic farming) to distinguish itself on the market, and hence to justify the price premium its survival requires: after all, why buy organic if it all contains GM material anyway? And, secondly, organic agriculture is arguably *more* than a simple economic choice. Organic agriculture provides a range of social benefits: environmental benefits most obviously, but also social and rural development, animal welfare benefits and, most controversially, public health benefits.[42] In these circumstances, the ability of organic production to distinguish itself from conventional or GM agriculture is of social importance. Chapter 4 returns to this subject.

We should also explore here some of the more autonomous ethical objections to GM agriculture. This cannot be anything like a full analysis, but some discussion is necessary if we are to examine the considerability of ethical concerns in regulation. Safety questions of course raise ethical issues, for example the ethical acceptability of risk to the environment or the health of producers, or responsibilities to future generations. Whilst some would argue

---

40 Ulrich Beck, *Risk Society: Towards a New Modernity* (Sage Publications, 1992).

41 See further Chapters 4 and 5.

42 These are benefits discussed in European Commission, Staff Working Document, *European Action Plan for Organic Food and Farming* SEC (2004) 739; see section 1.4.

that these are the *only* meaningful ethical issues at stake in GM technology, it is clear that there are other public responses to GMOs that might usefully be categorised in this camp.

A range of perspectives come together in concerns about setting limits for human ingenuity. The 'biocentric' ethical position identifies the cause of our environmental crisis as lying in a misunderstanding of the human place in nature. The dualist understanding of the world, with people separate from nature, is understood to create an illusion that human beings can conquer nature, and that further technological development and control of nature will improve the world. Biotechnology is then just the latest and most extreme example of the desire to control and subjugate nature.[43] The fear has to be that we compound our problems, or at best for anyone who believes in environmental limits, we simply delay the moment of reckoning. Similar concerns about the moral status of nature, although from a very different starting point, are expressed by the more common reference to misconceived attempts to 'play god' through technology.[44] This is apparent also in those parts of the popular press that have branded GM food 'Frankenfood'. Following his assertion that he is happy to eat GM food, and indeed that he gives it to his children, Tony Blair, the then UK Labour Prime Minister, was pictured on the front of the *Daily Mirror* (traditionally a Labour supporting newspaper) as a Frankenstein's monster, under the headline 'The Prime Monster'.[45] Whilst the 'Frankenfood' rhetoric has been widely dismissed as scare mongering and unhelpful,[46] the Frankenstein story may be a secular take on the 'playing god' idea, reflecting an instinctive anxiety about the consequences of human overreaching. When

---

43 Peter Bunyard and Fern Morgan-Grenville, *The Green Alternative* (Methuen, 1987), quoted in Andrew Dobson, 'Biocentrism and Genetic Engineering' (1995) 4 *Environmental Values* 3, pp. 231–6.

44 Georgiana Kirkham, '"Playing God" and "Vexing Nature": A Cultural Perspective' (2006) 15 *Environmental Values* 173.

45 *The Mirror* 16 February 1999. See the discussion in Alan Irwin, 'Constructing the Scientific Citizen: Science and Democracy in the Biosciences' (2001) 10 *Public Understanding of Science* 1. See John S. Applegate's discussion of the Frankenstein story, 'The Prometheus Principle: Using the Precautionary Principle to Harmonize the Regulation of Genetically Modified Organisms' (2001) 9 *Indiana Journal of Global Legal Studies* 207.

46 As recently as the end of 2006, the UK's Better Regulation Commission, *Risk, Responsibility and Regulation – Whose Risk is it Anyway?* (2006), approached the issue thus: 'the headline "Frankenstein foods" made it virtually impossible to have a balanced, evidence-based consideration of the advantages and disadvantages of genetically-modified crops in the UK. The headline stuck in people's minds and steered the debate in one direction only. A calmer consideration may have yielded a targeted regulatory option that could have dealt with the scientific uncertainties and public fears while retaining the possibility of commercial development for the UK', p. 17.

we put this alongside the high profile of ignorance and uncertainty in this area, concern about the consequences of institutional or scientific hubris has a substantive basis.

This book concentrates primarily on GM crops, but when we think about GM animals ethical issues may apply with extra force. This is not simply about degrees of animal welfare, complicated by the possibility of engineering animals with reduced capacity for suffering.[47] The 'intrinsic value' of a 'natural animal' makes concerns about transgressing boundaries much more visible when animals rather than plants are modified, for example in the creation of pigs that produce less noxious effluent, lower-fat meat, or organs for transplantation:

> What will it mean to come across a rabbit in the woods once genetically engineered 'rabbits' are widespread? Why would we have any more reverence or affection for such a rabbit than we would for a Coke bottle?[48]

While this leaves open all sorts of questions about our ability to develop these sorts of moral sensibilities, and indeed questions about existing reverence for rabbits and all they represent, it raises the possibility of meaningful losses. And, whilst at the moment we are moving only single or small groups of genes, the creation of a 'chimera' would exacerbate these issues still further. In the absence of imminent commercialisation, modification of animals has had a relatively low profile in comparison with the outrage about the actual modification of crops. Whilst the cloning of Dolly the sheep and her sisters had a high profile, this was probably at least as much because it made urgent the consideration of cloning of human beings. Whether and under what conditions society will be prepared to accept development of GM animals is still to be worked through.

From whichever direction we approach these ethical concerns, we arrive in the contentious area of the 'natural' versus the 'unnatural'.[49] It is difficult to make sense of a natural/unnatural distinction, and it should go without saying that all sorts of injustice have been the result of a distinction between natural and unnatural behaviour, or an assertion of limits set by god. Certain extreme

---

47 The distinction between animal rights and animal welfare becomes more urgent in the case of reduced sentience: 'And the animal welfare freaks won't be able to say a word, because this thing feels no pain' – of ChickieNobs, Margaret Atwood, *Oryx and Crake* (Virago, 2003), p. 238. Animal welfare is a factor in EU GMO regulation; see Chapter 3.

48 Bill McKibben, *The End of Nature: Humanity, Climate Change and the Natural World* (Bloomsbury, 2003), p. 230.

49 See the discussion of the Nuffield Council, above n. 16, especially paras 1.39–40.

forms of biocentrism can be uncomfortably misanthropic. Moreover, traditional farming involves the identification and manipulation of preferred traits in animals and plants, without any necessary regard for the integrity or value of the organism itself. And, although some natural limits that agricultural biotechnology can escape, particularly the ability of different species to breed, in the past had to be respected, the precise location of the boundary is elusive.

The ethical issues around proper limits to human ingenuity have been more closely examined with respect to *medical* biotechnology, for example human cloning or stem cell research. Perhaps the concerns are simply more compelling in respect of the application of technology to humans, but there are parallels with farming.[50] Just as some see human identity as at stake in questions of human biotechnology,[51] some see the natural world (and its identity) as fundamentally compromised by biotechnology. Bill McKibben (talking about human-induced climate change) argues that now that human beings have become so large in our numbers and impact that we affect everything, even the weather, we witness the end of nature as an autonomous entity.[52] The loss expressed here is compelling in its seriousness. Even if we find it difficult to empathise with McKibben's almost spiritual approach to the natural environment, or just find these emotions an unhelpful approach to policy, it does not seem appropriate (or even necessary) to dismiss this type of concern out of hand.

It is, however, an enormous challenge to come up with a way to take the various spiritual/religious concerns about biotechnology seriously. If the depth and complexity of intuitions about limits to technology are not acknowledged, we accept that any technological development is permissible – in effect, that what we can do we should do. But, at the other extreme, there is a danger in simply throwing up our hands and allowing the most morally conservative sections of society to dominate decision making, banning on those grounds the products of genetic modification. Andrew Dobson's approach is useful. He

---

50 See the interesting discussion of the 'bifurcation' of institutional consideration of agricultural and medical biotechnology in the US, Adam D. Sheingate, 'Promotion versus Precaution: the Evolution of Biotechnology Policy in the United States' (2006) 36 *British Journal of Political Science* 243. In the UK and EU as well, the medical and agricultural applications of biotechnology went down different regulatory routes, although in the EU in particular there is an effort at an integrated policy approach – perhaps in part to allow other approaches to biotechnology to 'borrow' the legitimacy of medical advances.

51 Jurgen Habermas, *The Future of Human Nature* (Polity, 2003); Francis Fukayama, *Our Posthuman Future: Consequences of the Biotechnology Revolution* (Profile Books, 2002); McKibben, above n. 19.

52 McKibben, above n. 48. The separateness of nature in this analysis takes us away from strict biocentrism.

canvasses the possibility that questions of biocentric value should be part of the assessment of an application of biotechnology, but without precluding the development of biotechnology:

> a case-by-case examination is appropriate in which the guideline questions would be *what* is being done to *whom* (or to what), and *why*. Biocentrics might merely demand in this context that the moral considerability of species . . . be taken into account as a further factor in the decision-making process.[53]

This is akin to Habermas's thoughts (in respect of human biotechnology) on 'keeping one's distance from religion, without closing one's mind to the perspective it offers'.[54] The saddening that results from living in an altered, 'unnatural' world may not be thought adequate, or indeed sufficiently distinct from what has gone before, to merit rejecting an entire technology. That does not mean that it should be unaddressed, irrelevant in decision making. Moreover, if addressed directly, it is likely that some ethical positions could be balanced by any real benefits to new applications of GM technology. Intuitive assessments of whether the cost is worthwhile can apply even if the cost is in our relationship with nature, or playing god (or Frankenstein). This is an ultimately anthropocentric approach, and will not satisfy the true biocentric, as it assumes that human wants and needs ultimately trump natural integrity. Nevertheless, it at least takes seriously the concern, shared with biocentrics, for the integrity of our relationship with nature, and brings into the open the question of what this technology means, and what it is for.

Conclusions: the Purposes of GMOs

There is little real consensus on what the regulation of GMOs should be about. An enormous range of issues seek a voice in the regulatory process. Certainly, safety for the environment and human health are important values and should be and are an important part of the regulatory framework. But the regulatory framework needs to go beyond safety and risk and engage properly with a fuller range of questions. As we see throughout this book, this is to a significant degree accepted in the mainstream approach to decision making. But the mere articulation of different perspectives is not enough for them to be heard, and that is proving far more difficult. The radical uncertainties and ignorance around new technologies demand greater interrogation of the purposes of GMOs, individually and in general, so that those who bear the uncertainties know why.

---

53 Dobson, above n. 43. Note that Dobson does discuss in this article the controversy over the existence of 'species'.

54 Habermas, above n. 51, p. 113.

Official and supportive discourse on GMOs is grasping the importance of purpose. The front page of the Commission's biotechnology website is worth citing at length:

> There are currently over 6.4 billion people living on the planet, a figure which is increasing by 77 million each year . . . today over 1.2 billion people, mainly women and children, are living in extreme poverty. Coping with this future population increase will pose severe social and environment challenges for global leaders, not least of which will be providing enough food to go round.
>
> . . . New biotechnology techniques have the potential to deliver improved food quality and environmental benefits through agronomically enhanced crops. Enhanced food and feed quality may be linked to disease prevention, and may result in the reduced use of chemical pesticides, fertilisers and drugs, leading to more sustainable agricultural practices in both the developed and developing world. Advances in biotechnology can also result in major health care benefits, allowing for the production of cheaper, safer drugs in large quantities . . .[55]

This is an interesting acknowledgement that what a technology is *for* does matter, although it is also a fairly crude effort to change minds. These promises of agricultural biotechnology are presented as if they were straightforward scientific fact, rather than hope and expectation. The controllability of the world is assumed, and the potential of molecular biology makes it seem suddenly possible to rethink profound social problems as if they were technical or scientific problems: a technical solution to poverty, lost crops, environmental degradation, climate change. This is also profoundly uncertain and value laden, and tells us little about how we should respond to the current generation of commercially driven GMOs.

THE HEGEMONY OF RISK?

Regulation (in the narrow sense of authorisation decisions) of GMOs concentrates (although not exclusively) on questions of 'risk', particularly risk to the environment and to human and animal health.[56] The discussion in the previous

---

55 http://ec.europa.eu/environment/biotechnology/index_en.htm (accessed December 2007). The webpage goes on to mention 'the need for responsible policies at EU and international level to ensure . . . concerns are addressed and that the protection of the environment and human health remains a priority at all times'.

56 The regulation is examined in Chapter 3. Elizabeth Fisher, 'Risk and Environmental Law: A Beginner's Guide' in Benjamin Richardson and Stepan Woods (eds), *Environmental Law for Sustainability* (Hart Publishing, 2006), discusses the way in which environmental regulation in particular turned into a question of 'risk' in the last two decades of the 20th century.

section should make clear the inadequacy of any regulatory framework that addressed only these questions. It should also be clear that, even in respect of health and the environment, scientific uncertainty massively complicates decision making on risk. There seems to be no end to the academic debate about the roles of and relationships between science and politics. Because, whilst science is not free of politics, and public views are not devoid of scientific underpinnings, there is a disjunction in the case of GMOs between public alarm and a mainly sanguine set of experts.

For many years, it was assumed by elites, governments and regulators that these sorts of disagreement rested on a combination of public ignorance of science and public irrationality when absolute safety cannot be assured. But a range of arguments and observations from the social sciences seem finally to have coalesced to convince decision makers, at least in principle, that non-experts have something positive to offer decision making. So, for example, longstanding observations that expertise arises in a particular historical, cultural and political context, and (sometimes imperceptibly and unconsciously) incorporates individual and collective value judgments, mean that the results of expertise are not the inevitable and objective 'sound science' that is promised.[57] In this context, alternative individual and collective value judgments should be heard. And increasing awareness of scientific uncertainty makes it much more difficult to use 'the facts' as the only rational basis for decisions. This is not to argue for relativity, or to dismiss the importance of expertise. But inevitable and inherent limitations to scientific analysis, such as the problems of ignorance and indeterminacy discussed above at pp. 29–30, are compounded by more banal errors, data gaps and necessary assumptions. An understanding of the fragility of the expert (scientific) view in the regulation of new technologies makes it difficult to dismiss out of hand competing public perspectives on risk. And those competing perspectives are not irrational, but are based on substance. For example, expert risk assessors may focus on numbers (lives lost/illness), whilst the public takes a multidimensional and qualitative approach, with particular hazards meaning different things to different people., depending on underlying values and the context of the risk.[58] So the broader concerns about agricultural biotechnology, including the distribution of risk and benefit or a view of the risk as externally imposed, will affect the level of risk perceived by the public, and most importantly the *acceptability* of that risk. As well as broadening the understanding

---

[57] Sheila Jasanoff's work is an important example of this approach, above n. 24; also, for example, *The Fifth Branch: Science Advisers as Policymakers* (Harvard University Press, 1990).

[58] See for example the work of Paul Slovic and his colleagues, such as Paul Slovic, *The Perception of Risk* (Earthscan, 2000).

of 'risk', these different approaches can challenge the very conceptualisation of the problem as one of risk. A complex social question such as what to do about agricultural biotechnology is likely to be misconceived if regulated only in terms of risk.

There are still respectable arguments in favour of keeping important and expensive matters of risk regulation away from the public,[59] but critiques of a wholly technical approach to 'risk', and recognition of the complexity of good decision making, are now in principle part of the mainstream in the EU, and also the UK.[60] The new orthodoxy emphasises a more open approach to decision making, and accepts in principle the political, contextual and value-based nature of regulation. A great deal of political attention has been paid to uncertainty in the scientific process, to public perceptions of risks and to public values that fall outside traditional scientific assessments. The evolution of policy thinking on risk cannot be properly explained by reference to any one event, but some of the intellectual and institutional roots of change can be found in the BSE crisis discussed in Chapter 1, which led to inescapable political awareness of uncertainty in the EU and its Member States. Periodic crises over risk issues and scientific information have continued to draw public attention to the fallibility of administrative decision making, and underlined its political sensitivity. The relationship between expertise and the public, between technocratic and popular decision making, is a significant challenge in the regulation of agricultural biotechnology.

Whilst science clearly cannot provide all of the information and judgments needed to make decisions on agricultural biotechnology, equally clearly we need the information that only it can provide. Ensuring ultimate political responsibility (and public involvement), alongside adequate scientific and technical advice, has proven difficult, despite increasing recognition of its necessity. The institutionalisation of a clear division between the decision-

---

59 See for example Stephen Breyer, *Breaking the Vicious Circle: Toward Effective Risk Regulation* (Harvard University Press, 1993); Cass R. Sunstein, *Risk and Reason: Safety, Law and the Environment* (Cambridge University Press, 2002), 'celebrates technocracy', but does subject technocracy to 'deliberative democracy'.

60 Alan Irwin, 'The Politics of Talk: Coming to Terms with the "New" Scientific Governance' (2006) 36 *Social Studies of Science* 299. See the material associated with European Commission, *White Paper on European Governance* COM (2001) 428 final. In the UK, see especially House of Lords Select Committee on Science and Technology, Session 2000–01, 3rd Report, *Science and Society* (although the Better Regulation Commission, above n. 46, is far more ambivalent, and arguably suggests something of a backlash). Of course this is not new everywhere! Some Member States, for example Denmark and the Netherlands, have been operating in an open and discursive way for some years, and may even be retreating; see Irwin, ibid. Barry Commoner, *Science and Survival* (The Viking Press, 1963) is an early argument on the topic.

making responsibility of political institutions and the advisory role of experts has been a common response to this dilemma. The EU has taken this division seriously, and in a number of contexts, most prominently for current purposes food, formalises the division of 'risk analysis' into separate stages of 'risk assessment', a technical process, and 'risk management', the political process during which final decisions are made.[61] This approach is pursued in the regulation of GMOs, as discussed in Chapter 3.

The recognition, through space for political 'risk management', that decisions on GMOs are politically significant is welcome. The distinction between the technical and the political is not, however, as clear cut as the risk assessment/risk management dichotomy might suggest. The dichotomy is really an effort to add a new alertness of the political nature of decision making onto old ways of doing things, without fully thinking through the implications. The uncertainties and implicit value judgments of the science/risk assessment stage may be under-examined, and then in turn the political judgment may be overly constrained by the 'advice'. In his detailed and subtle examination of the process leading up to the nuclear bombing of Hiroshima and Nagasaki, Jonathon Glover provides a chilling example.[62] Certain crucial issues, for example whether a harmless demonstration of the bomb would bring about swift Japanese surrender, were misconceived as technical questions (as questions of capacity), and delegated to a technical committee. The committee did not take a fully rounded view of the decision it faced, concentrating on technical aspects of the use of nuclear power. The political decision makers in turn saw themselves as constrained by technical advice: 'To look closely at this, one of the central decisions of the twentieth century, is to become aware of a moral vacuum'.[63] The context could not be more different, and the historical force of agricultural biotechnology remains to be seen. Even in obviously morally significant situations, however, the inappropriate conceptualisation of a decision as technologically determined, rather than value based, can impoverish decision making and allow for the evasion of responsibility.

The precautionary principle, which famously acknowledges the place of scientific uncertainties at the centre of decision making, is an important oppor-

---

[61] See European Commission, *Communication on the Precautionary Principle* COM (2000) 1 final; Regulation 178/2002 Laying Down the General Principles and Requirements of Food Law OJ 2002 L 31/1.

[62] Jonathon Glover, *Humanity: A Moral History of the Twentieth Century* (Pimlico, 2001), Chapter 12. Closer to home, Liz Fisher's discussion of the BSE crisis in the UK is interesting, raising the question of what the scientific advice was *for* – the providers of the advice and the recipients saw the advice very differently; see Elizabeth Fisher, *Risk: Regulation and Administrative Constitutionalism* (Hart Publishing, 2007), Chapter 2.

[63] Glover, above n. 62, p. 105.

tunity to open up space for broader perspectives on regulation. The precautionary principle provides, at its most basic, that scientific uncertainty does not in itself preclude regulatory action. Depending on the approach to the precautionary principle chosen, that regulatory action could be constrained by, for example, the severity or irreversibility of the risk, or by a cost–benefit analysis of regulatory action.[64] The precautionary principle is subject to vastly different interpretations and roles, and this is not really the place for a lengthy analysis of the competing philosophies of either risk or administration revealed by these debates. But the precautionary principle is an important principle of EU law, and hence also of the regulation of GMOs.

The precautionary principle applies broadly in the EU, to the regulation of environmental protection, human health and safety.[65] The European Court of Justice (ECJ) and the Court of First Instance (CFI) have both had a number of occasions on which to consider the precautionary principle, and the Commission has published an important Communication on the subject.[66] The precautionary principle is routinely taken to mean that 'where there is uncertainty as to the existence or extent of risks to [for example] human health, protective measures may be taken without having to wait until the reality and seriousness of those risks become fully apparent'.[67] The CFI has held that risk assessment is a necessary first step in the application of the precautionary principle, although by definition the risk assessment is not expected to provide certainty.[68] The Commission also characterises decision making in areas of

---

64 See for example *Rio Declaration on Environment and Development* (United Nations Conference on Environment and Development, 1992); European Commission, above n. 61.

65 Article 174EC applies the precautionary principle to the environmental title of the Treaty. Joined Cases T-74/00, T-76/00 and T-141/00 *Artegodan and Others v Commission* [2002] ECR II-4945 provide one of the decisions explaining the broader application of the principle. See discussion in Veerle Heyvaert, 'Guidance Without Constraint: Assessing the Impact of the Precautionary Principle on the European Community's Chemicals Policy' (2006) 6 *Yearbook of European Environmental Law* 27; Nicholas de Sadeleer, 'The Precautionary Principle in EC Health and Environmental Law' (2006) 12 *European Law Journal* 139.

66 European Commission, above n. 61.

67 Case C-241/01 *National Farmers' Union v Secretariat General du Gouvernement* [2002] ECR I-907; Case C-180/96 *UK v Commission* [1998] ECR I-3903; Case T-13/99 *Pfizer Animal Health SA v Council* [2002] ECR II-3305, para. 139. This is cited also in European Commission, above n. 61.

68 See especially *Pfizer*, above n. 67, and Case T-70/99 *Alpharma Inc v Council* [2002] ECR II-3495. These decisions were delivered on the same day. The decisions are almost identical, except that in *Alpharma* an expert committee had not been consulted, whilst in *Pfizer* the institutions had consulted the relevant committee but did not follow its advice.

environmental and public health protection as beginning with scientific risk assessment. The CFI distinguishes between a 'purely hypothetical' approach to risk and risk which is 'adequately backed up by the scientific data available at the time'.[69] The former may not be the subject of even precautionary regulation, but the latter is. This distinction is far from self-executing, illustrating clearly (in the words of Advocate General Mischo) 'the tension inherent in applying the precautionary principle: on the one hand, a measure cannot be based on a purely hypothetical risk, yet, on the other hand, one cannot wait until the risk has been established with certainty'.[70] Identifying a precise 'trigger point' for application of the precautionary principle is probably not possible, not least because this is context specific, depending on acceptability of risks in particular cases,[71] but it does not seem to have been set particularly high in the cases.[72] The language of 'plausible risk'[73] and 'reasonable scientific doubt'[74] are later efforts to find a way through this quagmire.

The precautionary principle can provide an important way to explain regulatory actions when uncertain science cannot. Inevitably, the precautionary principle does change the nature of the evidence required by the proponent of an activity, requiring at least a greater level of explanation than 'no evidence of risk'. The EU courts however seem also to go further and require the prioritisation of public over economic objectives: 'The protection of public health, which the contested regulation is intended to guarantee, must take precedence over economic considerations'.[75] This fits in with arguments that the precautionary principle, in the face of dominant economic explanations for policy, should be used to provide a connection with 'collective, democratic interests and the public domain'.[76] But the current operation of the precautionary principle seems not to be as absolute as this might suggest. The current batch of decisions are supportive of EU 'precautionary' action but, even though the

---

69 *Pfizer*, above n. 67, para. 144. European Commission, above n. 61, refers consistently to 'potential risk'.

70 Case C-192/01 *Commission v Denmark* [2003] ECR I-9693, para. 101.

71 Theofanis Christoforou, 'The Precautionary Principle and Democratizing Expertise: A European Legal Perspective' (2003) 30 *Science and Public Policy* 205.

72 Veerle Heyvaert, 'Facing the Consequences of the Precautionary Principle in European Community Law' (2006) 31 *European Law Review* 185.

73 *Denmark,* above n. 70, para. 102 (Advocate General's Opinion).

74 Case C-127/02 *Landelijke Vereniging tot Behoud van de Waddenzee v Staatssecretaris van Landbouw, Natuurbeheer en Visserij* [2004] ECR I-7405, para. 59.

75 *Pfizer*, above n. 67, para. 456.

76 Mike Feintuck, 'Precautionary Maybe, but What's the Principle? The Precautionary Principle, the Regulation of Risk, and the Public Domain' (2005) *Journal of Law and Society* 371, p. 372. Note that Feintuck rejects the process approach to the precautionary principle and argues strongly for a substantive content to the precautionary principle.

application of the precautionary principle is mandatory,[77] the broad discretion that the Courts grant the institutions makes it likely that the precautionary principle will rarely be successful in a challenge of an insufficiently precautionary decision.[78] Application in respect of Member State implementation of EU legislation might be more demanding. In a case about Member State obligations under EC nature conservation legislation, the ECJ has held that action at the point of 'reasonable scientific doubt' is mandatory.[79] The key to this will be in the application of 'reasonable' (although elsewhere in the decision the reference is simply to 'doubt'). This decision, moreover, was taken in a specific legislative context, and in particular we should note that the national imp-lementation of nature conservation legislation does not attract the same sort of internal market pressure as might products, including GMOs.[80] Internal market rules are discussed further in Chapter 3.

But more generally, the EU judiciary is unlikely to be intending a move towards a rather tricky interpretation of the precautionary principle, which posits that whenever there are threats to the environment or health the proponent of an activity must prove its safety, without reference to costs and benefits. Given that proof of 'no risk' is rarely if ever available, this 'absurdly strong'[81] approach to the precautionary principle could create equal and opposite risks to those created by the risky activity itself. As Cass Sunstein puts it, the problem with a strong approach to the precautionary principle is not that it 'leads in the wrong directions, but that it leads in no direction at all'.[82] Sunstein's argument is that the precautionary principle can only provide answers (of any description, right or wrong) because we fail to notice the risks of *not* developing a technology such as genetic modification. This is an important observation, although it should be placed alongside the frequent failure of regulators and developers to notice long-term, diffuse and uncertain risks, which the precautionary principle is supposed to address. The immediate benefits of certain technologies are easily visible, whilst the long-term and diffuse risks may be missed or under-emphasised.

---

77 *Artegodan*, above n. 65, para. 184.

78 Heyvaert, above n. 65. The Court considers whether there has been a 'manifest error or a misuse of powers or whether the Community institutions clearly exceeded the bounds of their discretion', for example *Pfizer*, above n. 68, para. 166.

79 *Waddenzee*, above n. 74

80 See de Sadeleer, above n. 65.

81 To borrow from the discussion of 'strong sustainability' in Herman E. Daly, 'On Wilfred Beckerman's Critique of Sustainable Development' (1995) 4 *Environmental Values* 49.

82 Cass R. Sunstein, 'Beyond the Precautionary Principle' (2003) 151 *University of Pennsylvania Law Review* 1003, p. 1054.

But the EU institutions, including the Courts, do not fall into the trap identified by Sunstein. Not only is cost–benefit analysis emphasised in the application of the precautionary principle,[83] but the search for 'zero risk' (which seems to mean here proof of no risk) is explicitly rejected. In its rejection of a 'purely hypothetical approach to the risk',[84] the CFI in *Pfizer* approves the parties' agreement that 'a "zero risk" does not exist, since it is not possible to prove scientifically that there is no current or future risk associated with the [activity]'.[85] As discussed above (p. 29), this makes sense to the extent that, if a failure to prove safety were necessarily to *preclude* a development, innovation would never be allowed. But the *Pfizer* approach rejects even the *possibility* of basing regulation on the inability to prove safety. Given that the benefits of the GMO in question may also be hypothetical or uncertain, this is a highly restrictive approach. The centrality of scientific evidence and risk assessment techniques means that the precautionary principle cannot be used to examine 'ignorance' – 'we don't know what we don't know'. Even in the application of the precautionary principle, law silences certain forms of uncertainty, on the assumption that uncertainty will be managed by science.

As noted above, the EU approach requires precautionary decisions to be 'adequately backed up by the scientific data available at the time'. However, scientific advisers are denied the final word because, whilst they are expert bodies, they have neither democratic legitimacy nor political responsibilities, and decisions on 'acceptable' risk are political. The Courts will, though, police the scientific credentials of the *political* decision. So a decision in which a political institution disregards the opinion of its scientific adviser must be accompanied by 'specific reasons for its findings', and the reasons provided by the decision maker 'must be of a scientific level at least commensurate with that of the opinion in question'.[86] Along similar lines, when EU institutions are

83 See for example *Pfizer*, above n. 67, paras 469 and following; European Commission, above n. 61, para. 6.3.1. Whilst cost–benefit analysis in this context is not necessarily a formal exercise of attributing values to different harms and benefits, the requirement that the pros and cons be weighed up reminds us that we cannot always assume a severe regulatory response in the face of the precautionary principle.

84 *Pfizer*, above n. 67, para. 143; Case T-392/02 *Solvay Pharmaceuticals BV v Council* [2003] ECR II-4555, para. 129.

85 *Pfizer*, above n. 67, para. 145; Commission, above n. 61. The phrase 'zero risk' is somewhat ambiguous – attempts to reduce the probability of an *identified* rather than hypothetical hazard to zero may be legitimate; see Case C-121/00 *Hahn* [2002] ECR I-9193 in respect of national measures; *Solvay*, above n. 84, in respect of Community measures. Alternatively, de Sadeleer, above n. 65, argues this just applies to the withdrawal of a product from the market, p. 163.

86 *Pfizer*, above n. 67, para. 199. Note the ECJ's rejection of a CFI decision for assuming that a scientific opinion should be followed, 'without explaining why the Commission was obliged to follow that opinion', especially given the existence of

required to assess complex facts of a technical or scientific nature, they can only adopt a preventive measure without consulting the relevant EU level scientific committee in 'exceptional situations', and where there are otherwise adequate guarantees of scientific objectivity.[87] Even if it is true that the way the CFI applies these conditions gives them little 'real bite', in terms of constraints on discretion,[88] they at the very least structure the form and the presentation of the decision, which must be scientific.

There are other interesting positions on the precautionary principle, which would take more seriously the many and profound manifestations of scientific uncertainty. If the precautionary principle implies a recognition that 'the facts' alone cannot be used to justify a decision, because the factual context is contingent, then alternative forms of legitimacy must be required.[89] Most commonly, the alternative legitimacy is sought in traditional forms of political legitimacy, which might include public participation in a decision. Alternative legitimacy is also sought in alternative forms of expertise. Below we will see how the incorporation of alternative fora of expertise, economic and ethical, have been used to bolster authority on GMOs in the UK.

The precautionary principle in the EU is impossible to pin down. It operates in a range of very different legal contexts – the two mentioned so far are scrutiny of EU measures and scrutiny of Member State implementation of EU law – and in different legislative frameworks. Moreover, the precautionary principle operates in the EU at the juncture (or perhaps maelstrom) of a number of legal and administrative contexts, embracing not only the 27 Member States but also the different bodies operating within the EU. And the application of the precautionary principle is highly context specific.[90]

We return to this case law in Chapter 3 (especially p. 80). For now, though, we can say that the EU approach to the precautionary principle emphasises the

---

other relevant scientific opinions, Case C-198/03 P *Commission v CEVA Santé Animale SA* [2005] ECR I-6357, para. 53.

87 *Alpharma*, above n. 68, para. 213.

88 Heyvaert, above n. 72. Note also Karl-Heinz Ladeur's observation of the degree of discretion left to administrative decision makers, 'The Introduction of the Precautionary Principle into EU Law: A Pyrrhic Victory for Environmental and Public Health Law? Decision Making under Conditions of Complexity in Multi-Level Political Systems' (2003) 30 *Common Market Law Review* 1455.

89 Elizabeth Fisher, 'Precaution, Precaution Everywhere: Developing a "Common Understanding" of the Precautionary Principle in the European Community' (2002) 9 *Maastricht Journal of International and Comparative Law* 7.

90 See Fisher, above n. 62, for discussion of the different contexts of the precautionary principle, and the importance of those contexts. In Chapter 3 below, the role of the internal market comes to the fore, and in Chapter 6 the impact of the World Trade Organization.

continued centrality of scientific risk assessment in 'precaution'. The Commission and the Courts insist that the precautionary principle must not be used to justify arbitrary or irrational decision making, assessed by the normal tools such as cost–benefit analysis and proportionality. But the high profile of the precautionary principle in the EU suggests that the logic of the precautionary principle's critique of scientific uncertainty has been accepted, and the EU interest in 'democratising expertise' suggests that it has been taken seriously.[91] Demanding reference to what science and expertise is available is no bad thing, but, in the ultimate return to science as the only legitimate basis for the final regulatory decision, there is a failure to grasp the profound implications of the limitations of science.[92] The EU approach to the precautionary principle attempts to manage uncertainty with the tools of science. This fundamentally misunderstands the real nature of uncertainty, which cannot always be resolved by more or better science and is too narrowly understood if thought to rest only on scientific disagreement. The return to science also undermines the regulation of non-safety issues. Debate around, for example, the commercial trajectory of GMOs is less compelling if we are under the illusion that decision making is based on objective and inevitable facts.

The EU's understanding of the precautionary principle is not on its own an effective tool with which to address the breadth of debate on agricultural biotechnology. And we should expect those with power to make use of legally entrenched opportunities. So the potential for disruption to food security raised by those who oppose agricultural biotechnology is likely to be matched by the potential of agricultural biotechnology to contribute to solutions for hunger under the same (precautionary) rubric. In short, regulatory reassurance that GMOs are subject to the precautionary principle falls far short of addressing the extent of the debate. The challenge is no longer in forcing legislators and regulators to accept the complexity of scientific evidence or the relevance of concerns not subject to scientific evidence. It is in making that acceptance mean something. The next section begins to explore this.

---

[91] As part of the 'governance' project, above n. 60, see Report of the Working Group, 'Democratising Expertise and Establishing Scientific Reference Systems' (2001), and European Commission, *Communication on the Collection and Use of Expertise by the Commission: Principles and Guidelines 'Improving the Knowledge Base for Better Policies'* COM (2002) 713 final.

[92] Joanne Scott and Ellen Vos, 'The Juridification of Uncertainty: Observations on the Ambivalence of the Precautionary Principle within the EU and the WTO' in Christian Joerges and Renaud Dehousse (eds), *Good Governance in Europe's Integrated Market* (Oxford University Press, 2002), observe that the precautionary principle might conceal rather than resolve the tension between expert and popular government.

DECISION MAKING ON GMOs: THE UK EXAMPLE

A purely scientific approach to risk assessment, indeed any sort of focus solely on questions of risk narrowly defined, would exclude a very wide range of concerns in the case of agricultural biotechnology. In the UK, science dominated the agenda on GMOs for many years and concern not based on scientific evidence of potential harm was dismissed as irrational, ignorant or at best irrelevant.[93] This narrow framework for decision making went along with a resistance to public involvement. The UK, however, known for its traditionally opaque approach to decision making in technical or scientific policy, was by 2005 described as 'the most active experiment station for the politics of biotechnology'.[94] UK agricultural biotechnology policy now recognises (at least in principle) the importance of public values and public participation, reflecting more general changes that both accept a more expansive approach to risk and move beyond a framing of GMOs solely in terms of risk. In 2003 the UK held an elaborate consultation on the commercialisation of GMOs, in response to public pressure, but more specifically to a proposal by a government advisory commission. The Agriculture and Environment Biotechnology Commission, an independent body set up precisely to provide strategic advice to government on the ethical and social issues associated with agricultural biotechnology, advocated a public debate that would 'expose, respect and embrace the differences of view which exist, rather than bury them'.[95] The public debate was filled out with two other strands: an economic study and a scientific review.

The government's Strategy Unit carried out a study into the overall costs and benefits of GM crops to the UK, resulting in a lengthy report, *Field Work: Weighing up the Costs and Benefits of GM Crops*.[96] Cost–benefit analysis is enormously controversial, along conceptual, ethical and practical lines. There is a danger that a concentration on economics in decision making can hide important uncertainties and value judgments behind a display of apparently inevitable numbers. Any attempt to simplify the complex issues surrounding GMOs behind a façade of objectivity would have been resisted and probably

---

93 See for example the discussion in Black, above n. 8; Matthew Kearnes et al., 'From Bio to Nano: Learning Lessons from the UK Agricultural Biotechnology Controversy' (2006) 15 *Science as Culture* 291.

94 Jasanoff, above n. 24, p. 9. For now we concentrate on the UK level, turning to the EU in the next chapter.

95 Agriculture and Environmental Biotechnology Commission, *Crops on Trial* (AEBC, 2001), para. 21. The AEBC was wound up in 2005.

96 Cabinet Office, *Field Work: Weighing up the Costs and Benefits of GM Crops*, available at http://www.cabinetoffice.gov.uk/strategy/publications.aspx (accessed December 2007).

counterproductive. And, in the context of the enormous uncertainty that we are dealing with here, there have to be questions about what can be achieved by attributing values to the various costs and benefits. The report is circumspect, emphasising the limited evidence and data on which it is based, and making 'no attempt to provide a single "net present value" of total costs and benefits'.[97] Instead, it paints a complex picture that emphasises the difficulty of the *political* judgments to be made, assessing the pros and cons of different regulatory approaches for the future of GM agriculture in the UK, including the possibility of 'shocks and surprises', and emphasising uncertainty.[98]

The Science Review was set up to consider the current state of the science behind GM issues, including clarifying the state of knowledge and areas of uncertainty. Its main novelty was in taking the questions for review from the public, examining the science on 'popular concerns and questions about GM crops and food'.[99] This reminds us how unhelpful it is to attempt to set up boundaries between science and politics, when it should be possible for them to inform each other. One of the important functions of the Science Review was to identify gaps in knowledge, looking more widely than case-by-case review under the regulatory provisions. There is clearly a danger that case-by-case review under the regulations (discussed in Chapter 3) will miss cumulative effects and unexpected interactions. The GM Science Review Panel published its final report in January 2004.[100] In respect of the many concerns identified, the Review asks whether there are 'gaps in our knowledge or scientific uncertainties' and whether these are significant. This is an important and innovative exercise, potentially responding directly to one of the major concerns about GMOs: uncertainty. But the answers provided to these questions are on the whole slightly disappointing. The Review identifies research necessary to fill gaps in knowledge. This is a useful first step, but one that addresses only a particular type of uncertainty, capable of being met by more or better research. Beyond this, the Review tries to reassure, focusing particularly on the existence of strong regulation. By way of example, in respect of the 'possible nutritional and toxicological differences in GM food', the Review closes with the following:

> wide-ranging regulations have evolved for GM crops over the last two decades and development will continue . . . in many respects there is far greater safety evaluation of GM crops and derived foods, which require extensive testing in comparison

---

97 Cabinet Office, above n. 96, para. 1.5.4.

98 Cabinet Office, above n. 96, Chapter 4.

99 GM Science Panel, above n. 22, p. 28.

100 GM Science Panel, *GM Science Review: Second Report: An Open Review of the Science Relevant to GM Crops and Food Based on Interests and Concerns of the Public,* available at http://www.gmsciencedebate.org.uk.

> with conventional crops, which often require no mandatory testing at all. As in all walks of life uncertainties exist but the benchmark for GM food is that it should be as safe as conventional food, which already has a history of safe use.[101]

The Report also repeatedly emphasises that uncertainty is the normal state of affairs. As well as the above, for example:

> Absolute certainty about lack of allergenicity cannot be achieved . . . in this or any other risk assessment. The likelihood that all regulatory and safety testing procedures fail is probably small but cannot be quantified at present as no data are available that allow us to do so. This, however, is not a unique situation in risk assessments. Absolute safety does not exist.[102]

Whilst the Science Review was innovative and progressive – an effort to move beyond some of the limitations of regulatory risk assessment in this area – the omnipresent and more complex issues of uncertainty were not really addressed. What to do about ignorance and indeterminacy is as much a social as a scientific problem, and so this is not wholly surprising in the work of scientists. But the Science Review repackages uncertainty as a manageable and routine risk. And the government response to the Review, if anything, emphasises this confident managerial approach to the problem of scientific uncertainty. Government considered the key conclusion of the Review to be that 'GM is not a single homogeneous technology', and so blanket judgments cannot be made.[103] In the words of the Secretary of State, the Review 'reported no verifiable ill-effects from extensive human and animal consumption of products from GM crops over 7 years, and it concluded too that current GM crops were very unlikely either to invade the countryside or to be toxic to wildlife'.[104] This is far from capturing the emphasis on uncertainty even in the Review itself, and reminds us of the possibility that politicians and scientists will expect too much from each other. And consistently with the Review, when looking at uncertainty, the government concentrates on the promise of new or recent research, as well as the existence of safety regulation.[105] So, for example, in respect of the uncertainty identified around the 'Impact of the use of broad spectrum herbicides that is associated with the cultivation of herbicide tolerant crops, including the implications throughout rotations and on higher

101 GM Science Panel, above n. 22, p. 78. There are many similar examples.

102 GM Science Panel, above n. 22, p. 89.

103 Department for Environment, Food and Rural Affairs, *The GM Dialogue: Government Response* (DEFRA, 2004).

104 Secretary of State Margaret Beckett's Statement on GM Policy, 9 March 2004.

105 DEFRA, above n. 103, annex 2.

trophic levels (e.g. farmland birds)', the government response is to refer to the 'Farm Scale Evaluations' (FSEs), which were scientific studies addressing the biodiversity impact of particular GM crops at the farm level.

Whilst not strictly part of the 2003 GM Dialogue, the results of the FSEs obviously met with interest. They were progressive in their assessment of the impact of the entire farm management regime associated with particular GM crops on biodiversity, rather than the GMO in isolation. The FSE results demonstrated that certain genetically modified herbicide-tolerant (GMHT) crops, with their associated herbicide regimes, resulted in lower levels of field biodiversity than was the case for their conventional counterparts managed conventionally. Another, by contrast, resulted in greater levels of field biodiversity than was the case for its conventional equivalent, conventionally managed.[106] Perhaps the most controversial element of the FSEs was the benchmark against which the GM crops were measured. Using conventional farming as the benchmark was described as 'unambitious'. And GMHT forage maize was compared with a conventional crop treated with a herbicide called atrazine, which was described in evidence to a parliamentary committee as a chemical that 'turns a maize field from what was once a diverse grass field . . . into a wildlife desert. It is really ground zero as far as wildlife is concerned'. The use of atrazine has now been phased out for environmental reasons, and this comparison was described by the parliamentary committee as 'invalid'.[107] GM agriculture is clearly going to look better in terms of biodiversity when judged against intensive farming than against, for example, organic farming.[108] The controversy over choice of comparator (conventional farming and atrazine) reminds us that the technical information provided to decision

---

[106] See ACRE, *Advice on the Implications of the Farm-Scale Evaluations of Genetically Modified Herbicide Tolerant Crops*, 13 January 2004. The results are only applicable to particular management regimes and so are of less relevance if future management techniques (either for the GM crop, or for the conventional comparator) change. The Department for Environment, Food and Rural Affairs has stated that, when seeking renewal of authorisations, new evidence providing a comparison with conventional practice current at that time will be required: DEFRA, *Government Response to the Environmental Audit Committee Report: GM Foods – Evaluating the Farm Scale Trials*, Second Report, Session 2003–04.

[107] Environmental Audit Committee, 2003–2004 Second Report, *GM Foods – Evaluating the Farm Scale Trials* HC 90-I, paras 6, 48 and 50.

[108] This comparator was accepted (apparently with no inhibitions) by the GM Science Panel, above n. 22. We should also recall the commonplace observation that farming is generally less well environmentally regulated than point source industrial pollution (e.g. factories). The comparator need not be farming at all. This is particularly pertinent if farming really is going to replace factories for the production of certain materials – bio-pharming must not be allowed to shift pollution from heavily regulated and monitored factories to weakly regulated and monitored fields.

makers is framed by a value-laden decision. In many cases, these values could be hidden by their apparently technical nature. In this case, however, the wide and heated discussion of the FSEs illustrates how transparency of scientific assessment can bring judgments and alternatives into the open: science as an open and political endeavour.

Described by its organisers as an 'unprecedented event – a special public debate before a potentially far-reaching change in public policy',[109] the public debate strand of the GM dialogue went under the slightly sinister title of *GM Nation?*. It constituted a self-consciously deliberative and inclusive exercise.[110] Whilst there have been criticisms of the process, this has generally been in the context of an acknowledgement of the experimental nature of the process, and of the financial and time limitations.[111] The core activity of *GM Nation?* was a series of public meetings around the country, including a few high-profile regional events, together with a larger number of smaller local meetings, from which written feedback was provided to the Steering Committee. The *GM Nation?* public meetings tended on the whole not to follow the traditional approach to public meetings in the UK. Rather than revolving around platform addresses from experts, followed by questions from the floor, and in line with the deliberative ethos of *GM Nation?*, attendees were encouraged to listen and engage with other opinions, as well as put forward their own views. *GM Nation?* was an interesting and innovative approach to public participation, moving beyond (although including) simpler forms of written consultation, allowing or encouraging engagement between the participants. The *GM Nation?* process swept up the 'other' issues that provoke concern about GMOs, as well as allowing lay discussion of scientific and economic issues. There appeared to be no attempt to impose a framework of 'acceptable' ways of discussing the 'problem'. Instead, the public framed the issues for the debate through discussion workshops.[112]

The *GM Nation?* debate demonstrates what can be done in terms of public involvement in complex policy areas. It also illustrates some of the difficulties with moves to more extensive public participation, which should not be greeted uncritically. Not least, participation absorbs time and financial

---

109 *GM Nation? The Findings of a Public Debate* (2003).

110 Its record in both of those respects is mixed, Understanding Risk Team, (T. Horlick-Jones et al.), *A Deliberative Future? An Independent Evaluation of the* GM Nation? *Public Debate about the Possible Commercialisation of Transgenic Crops in Britain, 2003* Understanding Risk Working Paper 04-02.

111 Understanding Risk Team, above n. 110; Environment, Food and Rural Affairs Select Committee, Session 2002–2003 18th Report, *Conduct of the GM Public Debate*, HC 1220.

112 *GM Nation? The Findings of a Public Debate*, paras 14–18.

resources, and the design of an effective system is hugely difficult. The question of *who* participates is always controversial, as the rhetoric of inclusion that goes along with participation is matched by the real dilemma of exclusion. Those poor in time or other resources (just poor perhaps), the uneducated or the inarticulate are likely to be excluded. In this case, there were clearly difficulties reaching out beyond the 'usual suspects' with a pre-existing interest in agricultural biotechnology. And the purposes of *GM Nation?*, as is often the case with exercises in public participation, were unclear, even retrospectively.[113] Government had undertaken to respond to the results of the GM Dialogue, and, although there was 'widespread cynicism' about the impact of the debate on decision making,[114] the 'dialogue' was unique for the UK. At the very least its political riskiness suggested a desire to engage with public debate on this issue. There was clearly no commitment to seek or follow a majority view on GMOs, but there equally clearly was an effort to make all parties feel that they have sufficient say in decision making. *GM Nation?* sought to bring into the fold a range of information that would be difficult to locate in a single bureaucracy, including external information on the economics and the science, but, more strikingly, information on public views.

Perhaps the most significant inherent difficulty with such a far-reaching public participation process, however, is quite what will be done with the results. In particular, there is a question as to how much freedom (legally, economically, politically) government has to respond to unwelcome views. Undeniably, the timing of the debate in this case was problematic, since considerable economic and organisational commitments were already in place. And such an ambitious programme was only possible at all because of the unanticipated regulatory hiatus prompted by the moratorium, itself the result of public pressure. Government made a joint response to the three strands of the dialogue. It claimed to 'take public concern very seriously' and to have 'weighed public opinion alongside the scientific evidence'.[115] It accepted that people are 'generally uneasy', and that there is 'little public support for early commercialisation'.[116] Indeed, the polarisation associated with GMOs was not particularly apparent. The government response acknowledged the 'complex range of issues and concerns' that shape peoples'

[113] Understanding Risk Team, above n. 110.

[114] Understanding Risk Team, above n. 110. The 'institutional body language' of government may have been problematic: a number of ministers had made clear their support of GM, and the Prime Minister had criticised anti-GM protestors before the debate got underway, p. 35.

[115] DEFRA, above n. 103, Executive Summary, para. 1.

[116] DEFRA, above n. 103, Executive Summary, para. 7.

views on biotechnology,[117] reinforcing its acknowledgement of the legitimacy of issues going beyond technical assessments of risk to health and the environment. The government claimed to have 'taken into account a range of different policy objectives: environmental protection, food safety, consumer choice, sustainable food and farming, thriving rural communities, science and innovation, industrial competitiveness, international development, and trade'. There was no suggestion that government had a purely technical or scientific decision to make. To borrow from Andrew Dobson in another context, we might be forgiven for thinking that government 'had fully grasped the normative dimension' of decision making on GMOs, 'and was determined to factor it into decision-making'.[118] The government response, however, turned very quickly to its commitment to 'evidence-based policy-making', for which we might read a far more familiar reliance on scientific evidence: 'the scientific evidence supports neither an outright ban nor a blanket acceptance of all GM crops'.[119]

As it went through 'each of the concerns raised in the public debate', the focus was on responding to those concerns through rigorous safety and approval processes. The concerns were listed as follows: (1) caution and precaution; (2) protecting human health; (3) protecting the environment; (4) providing choice for consumers; (5) providing information; (6) openness and transparency; (7) gaps in scientific knowledge; (8) developing countries; (9) no need for GM crops?; (10) ethical issues.[120] The response of government to the first six issues was basically to explain how both current regulatory practice and available scientific information, including that gained during the Dialogue, responded to these issues. So, for example, on concerns about human health, the government response was as follows:

> The science review concluded that there is no evidence to suggest that current GM foods pose a greater risk to human health than their conventional counterparts. All GM food and animal feed is strictly regulated in the EU and is subject to a comprehensive safety assessment. The EU approval process has recently been further strengthened and we believe that it is sufficiently rigorous to ensure that approved GM foods are as safe as their non-GM counterparts.[121]

Public concern was met primarily with reassurance, by reference to scientific evidence and its use by regulators. Scientific values were emphasised, and the possibility and desirability of controlling our environment was assumed. The

---

[117] DEFRA, above n. 103, Chapter 3, para. 3.1.

[118] Andrew Dobson, *Citizenship and the Environment* (Oxford University Press, 2003), p. 153.

[119] DEFRA, above n. 103, para. 11 (Executive Summary), and para. 4.4.

[120] DEFRA, above n. 103, Chapter 5 (numbering added).

[121] DEFRA, above n. 103, para. 16.

last four criteria are potentially more challenging, not obviously susceptible to this type of reassurance. Nevertheless, the government response was not ambitious. As discussed above, when addressing concern about gaps in scientific knowledge, the government acknowledged uncertainty and promised further research. This faith in more or better science slightly misses the point. The most profound dilemmas around ignorance (we don't know what we don't know) and indeterminacy (the unpredictability of real-life ecological and human systems) were not addressed, and yet lack of knowledge was a key concern in *GM Nation?*. On the question of developing countries, government asserted the value of developing countries making their own decisions on GM crops. This is clearly true, but disingenuous in its silence on the influence of EU policy.

The public debate identified the 'need for GM crops' as a significant issue. This is an enormously important and complex question, feeding into the appropriateness of bearing uncertainty and the purpose and role of agricultural biotechnology. The government dealt with this very briefly, citing immediately the Economic Review to the effect that, whilst there is currently limited economic value to the UK in GM crops, they have the potential to offer greater benefits. This silences the circumspection of the Economic Review. And, rather than engaging with the underlying question of what the technology is really for, the government responded to this question from a wholly economic perspective, even to the extent of saying that 'ultimately the market will decide'.[122] Again, this rather misses the point of who benefits from GMOs and how, and is based on unacknowledged uncertainties, not to mention unstated values. And the final, and related, question of ethical issues was met by reference to the Nuffield Council on Bioethics, to the effect that 'there is an ethical obligation to explore [the] potential benefits [of agricultural biotechnology] responsibly, in order to contribute to the reduction of poverty, and to improve food security and profitable agriculture in developing countries'. The Nuffield Council on Bioethics is part of a charitable trust, with no formal authority but considerable informal influence. In the report cited by the government, it dismissed the notion that ethical concerns should limit development of biotechnology, beyond questions of safety (that is, the ethical implications of harming the environment or the health of farmers).[123] It took a utilitarian approach to weighing the costs and benefits of GMOs, and dismissed any attempt to distinguish between forms of agriculture on the basis of their 'naturalness'. This utilitarian approach assumes that we have a better understanding of the effects of GMOs than we do, shackling the ethical debate to the

122 DEFRA, above n. 103, para. 5.7. See Chapter 4 for a discussion of labelling.
123 Nuffield Council, above n. 16.

scientific evidence. The Nuffield Council addressed also the argument that 'industrialisation' of farming is 'culturally impoverishing', leading us to 'lose touch' with the 'sense of things', but determined that this provides 'little justification in banning GM crops . . . when the rest of society travels so substantially in [this] direction'.[124] The overlap between expertise and implicit values and assumptions is especially pronounced.

The appeal of 'expertise', when faced with the prospect of having to engage with competing perspectives on new technology, is very clear. The effort to sidestep or defer, or sometimes even resolve, political disagreement by delegation to experts is very familiar. And the experts relied on here are not just experts in science and technology. So the economic experts are presumed to have an authoritative view on the need for GMOs. Whilst economic information has a great deal to offer, this closes out alternative understandings of progress, and assumes that the dominance of commercial objectives in the pursuit of this technology is unproblematic. And ethical experts are deemed to provide a response to ethical dispute. Whilst, of course, there is much to learn from those practised in a particular form of discourse, there is a danger that the professionalisation of ethics could put ethics (like 'risk') beyond the reach of popular politics or ordinary political deliberation, disempowering the ordinary language of politics and further constraining political decision making. The approach of government minimises the value assumptions implicit in its own policy, whilst emphasising the values implicit in the perspectives of those outside the official discourse. Everyday intuitions on social and ethical impacts of GMOs may be rejected or may be wrong, but should not be silenced.

The government ostensibly accepted the wide range of issues that are at stake in the regulation of GMOs. However, the rhetorical responsiveness to a full range of considerations proved difficult to match with the actual practice of decision making, where government has tended to go back to expert advice and scientific information. Strides were taken in the provision of institutions for open decision making, for cooperative and communicative regulation. However, the weight of the different perspectives clearly varies, and certain voices continue to find a place in the debate more easily than others. This provides a useful case study of both the potential for progressive approaches to assessing biotechnology and the need to look behind what decision makers say they are doing. Later chapters examine some of the legal pressures that point towards the narrowing of the basis for decisions. There are also political pressures. Science remains a very powerful legitimising force, with an air of objectivity and inevitability that cannot be found in political judgments.

---

124 Nuffield Council, above n. 16, para. 1.43.

The far-reaching claims made for agricultural biotechnology remind us that scientific endeavour is not isolated from social and ethical expectation. Social scientists talk about the 'imaginaries', the projections and expectations, associated with new technologies,[125] the visions of social transformation through science that are bound up with basic research. Such imaginaries have considerable power, but neither their realism nor their desirability is subject to much discussion at the early stage of 'basic' research. And, at the later stage, there is a danger that the search for a technical solution to the problem identified is a fait accompli. So, for example, GMOs might be associated with increasing agricultural production of staple crops in parts of the world where hunger is a problem. At the early stages, there is no external questioning of the likely trajectory of actual research, nor whether this would be the best way to tackle poverty. A social, economic and political problem (extreme poverty) is hence reconceptualised as a technical problem.[126] This can obscure possible alternative solutions. But, in addition, the risks and feasibility of the technical solution become the central question, and it is natural to turn to the technical experts for an answer.[127] This is related to the very long and honourable tradition of progress through science and technology, to which profound concerns around biotechnology pose a genuine challenge. Where science cannot provide sufficient authority for a particular stance, alternative sources of authority are sought. Here, those alternative sources of authority are economic and ethical expertise, which share some of the air of objectivity and independence of science.

'Public opinion' should not of course be followed in all cases: there is rarely even a monolithic public opinion *to* follow, and in any event public opinion can be manipulated, difficult to identify, and discriminatory, ignorant or ill-informed. But, just as it is rhetorically accepted that the scientific information is not decisive, so religious or ethical inputs can be significant without being decisive. An open engagement with the underlying basis for public concern (however difficult and controversial the identification of that underlying basis inevitably is) would have a number of benefits, aside from the virtue of democratic responsiveness. It would put into practice the basic acceptance that policy on GMOs is not a wholly scientific question, but one resting

---

[125] See the discussion in Kearnes et al., above n. 93. See also the discussion of 'visions' of nanotechnology in Royal Society and Royal Academy of Engineering, *Nanoscience and Nanotechnologies: Opportunities and Uncertainties* (2004).

[126] See the discussion in Kearnes et al., above n. 93; see also Habermas, above n. 51.

[127] Sagoff, above n. 17, provides a nice illustration: 'How can Malthusians plausibly argue that genetic engineering cannot achieve these breakthroughs? Only those well versed in the technology itself can speak to its possibilities, and they are full of optimism', p. 16.

on fundamental values about how we wish to live. If nothing else, religion and spiritual environmentalism have a powerful language for the uneasiness that surrounds certain new technologies. Government could ultimately prefer a particular scientific view, but on the basis of an explanation that, for example, in this case the dangers of ignorance are worth running, as they have been with earlier technological developments. Neither science nor moral absolutes can spare decision makers from the decision. Equally importantly, engagement with the basis of public concern would allow the considerations on which a decision is based to be challenged by competing positions in a way that is not possible if a political decision is justified by a convenient scientific framework. This in turn provides a level of 'check' on the risk of succumbing to irrationality and prejudice.

CONCLUSIONS

The promised commercialisation of GMOs has raised new and challenging dilemmas for law and regulation. Subsequent chapters explore the legal response to GMOs in a number of fora. The purpose of this chapter has been to study the nature of the debate on GMOs, to allow an examination later of how that debate resonates within the legal framework for decision making. Some see agricultural biotechnology as a simple next step in the progress of agriculture, no different in kind from all other technical advances. Others see it as a qualitative change, altering our relationship with nature in profound and unfamiliar ways. Supporters of agricultural biotechnology promise environmental benefits, food for the hungry, a way to live with climate change, improved health, cleaner manufacturing and prosperity. Opponents are angry at the lack of attention paid to the possibility of environmental disaster, unpredictable risks to human health, impoverishment of all but large corporate investors and an end to our relationship with nature as we know it.

The breadth of concern places this debate beyond its own context of what we should do about GMOs, and firmly in the context of the appropriate path of industrial development and the appropriate role of technology. In the early days, the only acceptable terms of debate were narrowly scientific, and focused on provable negative impacts on, primarily, human or animal health and the environment. There are really two concerns about the reliance on scientific expertise, partially contradictory, but fundamentally connected, and going to the heart of the transparency and accountability of decision making. First, it is possible that political decisions made to respond to public opinion will be 'dressed up' as scientific decisions. Secondly, the depth and extent of public concern will not receive a full response. It is increasingly accepted that science does not have all of the answers and that 'risk' does not capture all of the story. This goes along with an acceptance that values matter to regulation and that the public has something to contribute to decision making. But questions remain about the extent to which law is capable of recognising and accommodating this broader framework. Sheila Jasanoff's identification of the dual nature of 'representation' is telling: 'It refers both to the self-presentation *of* the public within and before governing institutions, and the presentation *by* the public to those bodies of matters that are seen to be of collective significance'.[128] It is the latter that seems to be most testing. Whilst there are many possible roles for 'public participation', an important one for current purposes is linked with the need to explore this massive scope of issues raised by GMOs.

The 'why' question is crucial in the regulation of GMOs, as is the question of who owns and who benefits from this technology. We should also acknowledge that the unknowns go far beyond the scientific questions. Uncertainties extend to, for example, the impact of biotechnology on small farmers and its impact on the corporate role in the food sector. There are some powerful arguments for pushing ahead with GMOs in farming. Regret for the path not taken is always likely, and we should not think that there are easy answers. How these questions resonate in the regulatory framework applying to GMOs is a topic to which we return throughout this book.

128 Jasanoff, above n. 24, p. 281.

3. GMOs and risk regulation in the EU

INTRODUCTION

An economic understanding of biotechnology pervades the EU's *Life Sciences Strategy*, pointing towards the advantages of speedy commercialisation of GMOs.[1] This perspective is if anything increasingly emphasised in the ongoing annual reviews of the Strategy: the overriding objective seems to be 'to improve the situation for European biotechnology'.[2] Whilst the contribution of biotechnology to economic development tends to be presented as objective fact, a necessity to which we must adapt, these conclusions are profoundly value laden. One of the important implications of conceptualising biotechnology as an economic question is to put biotechnology into terms of competition with the US (for example). This contributes to the presentation of biotechnology as a *European* problem, which should be solved for a *European* public.[3] It is important to remember that it is not just the opponents of agricultural biotechnology who have social and ethical commitments on the topic. The social and ethical commitments of the policy framework are often silently taken for granted.

---

1 European Commission, *Life Sciences and Biotechnology – A Strategy for Europe* COM (2002) 27 final (the *Life Sciences Strategy)*, p. 7. The *Strategy* was drafted by the Commission and 'welcomed' by the Council; see *Council Conclusions and Roadmap of 26 November 2002 for a Strategy on Life Sciences and Biotechnology* [2003] OJ C 39/9. Other arguments in favour of pursuing agricultural biotechnology are also part of EU policy; see Chapter 2.

2 COM (2004) 250 final, p. 2. The *Life Sciences Strategy* has been subject to annual 'progress reports'; see COM (2003) 96 final, COM (2004) 250 final, COM (2005) 286; see also European Commission, *Communication on the Mid-term Review of the Strategy on Life Sciences and Biotechnology* COM (2007) 175 final: 'there is a strong need to continue promoting the development of life sciences and biotechnology in the EU', section 6.

3 Sheila Jasanoff, *Designs on Nature: Science and Democracy in Europe and the United States* (Princeton University Press, 2005). See *Life Sciences Strategy,* Chapter 5. This approach can be traced back to at least European Commission, *Biotechnology in the Community* COM (1983) final, which refers to competition with Japan and fragmentation of the research effort. Joint Research Centre Reference Report, *Consequences, Opportunities and Challenges of Modern Biotechnology for Europe* (JRC European Commission, 2007) makes constant comparison to the US, and to a lesser degree, Japan.

But, as well as seeking to encourage and support a buoyant market in biotechnology, the EU imposes relatively burdensome regulatory obligations on GMOs, primarily along the parameters of risk to the environment and to human health. This fits in with the EU's generally enhanced role in 'risk regulation' over recent years, and again points towards a Europeanisation of the policy area. Risk regulation is in part about maintaining the integrity of the internal market, but also rests on a sense that successful EU risk regulation can contribute to the political legitimacy and popular relevance of the EU, by providing a visibly effective response to public concerns. Food in particular has often shown itself to be both particularly challenging to the internal market and particularly sensitive in its public profile,[4] and the food use of GMOs is prominent in the EU regulation. The periodic controversies over food risks in the EU are also controversies about the modern role of agriculture. Agriculture has long been a central plank of EU policy, and historically played a very significant role in EU 'polity building'.

Along the different parameters of both seeking benefits and controlling risks, GMOs are a thoroughly Europeanised policy area. The deliberate release into the environment of GMOs has required authorisation throughout the EU since 1990.[5] This early legislation was the subject of bitter complaint from the industry. In particular, by contrast with the US, the EU concentrated on the *process* of genetic modification, rather than the end product. The latter approach suggests lighter regulation, as it implies that the safety of all products (so all tomatoes, regardless of how they are produced) should be assessed on the same criteria. Looking at the *process* involves distinguishing between products and regulating GM as a special technology, and was always likely to lead to more intensive regulation. Nevertheless, over time, the 1990 Directive was perceived to be deeply inadequate for the regulation of agricultural biotechnology. As discussed in Chapter 1, in 1999 12 Member States declared their opposition to further authorisations of GMOs,[6] inaugurating the '*de facto* moratorium' on new authorisations. No decisions, positive or negative, were made on applications between 1998 and 2004, and a number of Member States instituted safeguard measures of doubtful legality, banning from their territory GMOs that had already been approved at EU level. Whatever one's view on

---

4 Damian Chalmers, '"Food for Thought": Reconciling European Risks and Traditional Ways of Life' (2003) 66 *Modern Law Review* 532.

5 Directive 1990/220 on the Deliberate Release into the Environment of Genetically Modified Organisms [1990] OJ L 117/15.

6 Two declarations were made by different groups of Member States in the 2194th Council Meeting – 24/25 June 1999: *Declaration by the Danish, Greek, French, Italian and Luxembourg Delegations Concerning the Suspension of New GMO Authorisations*; *Declaration by the Austrian, Belgian, Finnish, German, Netherlands, Spanish and Swedish Delegations*.

the technology, the breakdown in EU law brought about by this notorious episode was an extraordinary, indeed unprecedented, event. The Commission, probably wisely, did not immediately seek to force the issue, and along with the other institutions (including the Member States in Council) used the moratorium as a period for the negotiation of new legislation. The concerned Member States had demanded stricter risk assessment, and emphasised the need for rules on labelling, traceability and liability. To a greater or lesser degree, all of those issues received legislative attention by 2004. A complex set of legislation now attempts to respond to the perceived inadequacies of the earlier legislation, and to divide sensitive responsibilities on GMOs between national and EU bodies, and between scientific, political and indeed market[7] authority. The authorisation of GMOs is the subject of this chapter. Following an outline of the regime, there is an effort to account more fully for where authority lies on this topic. The role of scientific risk assessment and of political considerations beyond risk assessment will be examined, before turning to the very complex relationship between central and national levels of governance.

The moratorium arose out of conflict and chaos. Although it did provide a breathing space for the renegotiation of the regulatory framework applying to GMOs, there was never any clear and unified EU decision to that end. The EU remains far from speaking with one voice on agricultural biotechnology, and the ending of the moratorium under this new legislative regime has been just as conflictual as its inception. Draft authorisation decisions are drawn up by the Commission and sent to the Council for its consideration. Council can reject or accept the Commission's draft by qualified majority voting, but in the absence of a Council decision the Commission adopts its draft. All of the authorisations from the ending of the moratorium in 2004 to date (December 2007) came from the Commission following Member State failure in Council to reach a qualified majority in either direction. Legally, the Commission is obliged to see the process through in this way: if the Council fails to act, 'the proposed implementing act shall be adopted by the Commission'.[8] This avoids regulatory impasse, but the political stakes are high, and those cases where

---

7 After authorisation, GMOs must be labelled and traceable; see Chapter 4.

8 Decision 1999/468 Laying Down the Procedures for the Exercise of Implementing Powers Conferred on the Commission [1999] OJ L 184/23, Article 5(6). This legislation is part of the EU's general administrative provisions rather than specific to GMOs. Decision 1999/468 has been amended by Decision 2006/512 [2006] OJ L 200/11, which introduces a new 'regulatory procedure with scrutiny' in respect of 'measures of general scope designed to amend non-essential elements of a basic instrument'. This procedure would allow the European Parliament, like the Council, to reject Commission proposals. The authorisation process is not likely to be affected, but other parts of the GMO regime are being changed, for example amending annexes, changing

qualified majorities cannot be reached are precisely the cases in which it is most problematic for the Commission to forge ahead without broad national support.

The EU is of course accustomed to reconciling different interests and perspectives, and a tolerable consensus (or at least provisional acceptance of majority views) is usually achieved. The tensions in agricultural biotechnology policy are, however, unusually persistent. In part, these tensions are inherent in moves to develop and adopt a new technology, torn between an eagerness to reap benefits, economic and social, and a desire to minimise disruption: 'newness . . . is almost always packaged with a stated capability for fulfilling dreams, and, simultaneously, challenging existing institutions and mores'.[9] The tension also, however, reflects fundamental divisions between the Member States on the appropriate role for biotechnology in agriculture. The ways in which law attempts to negotiate these tensions will be one of the main themes of this chapter.

THE AUTHORISATION REGIME

The first major legislative intervention in the renegotiation sparked by the moratorium was Directive 2001/18 on the Deliberate Release into the Environment of Genetically Modified Organisms (the 'Deliberate Release Directive').[10] The Deliberate Release Directive imposes an obligation to seek authorisation for the marketing of any GMO, including those that will be deliberately released into the environment.[11] The Deliberate Release Directive was almost immediately amended by Regulation 1829/2003 on Genetically

---

the information that needs to be provided with an application: see European Commission, *Proposal for a Directive amending Directive 2001/18 concerning the deliberate release into the environment of genetically modified organisms, as regards the implementing powers conferred to the Commission* COM (2006) 920 final; European Commission, *Proposal for a Regulation amending Regulation No 1829/2003 on genetically modified food and feed, as regards the implementing powers conferred on the Commission* COM (2006) 912 final.

9 Monroe E. Price, 'The Newness of New Technologies' (2001) *Cardozo Law Review* 1886 (discussing information technologies).

10 Directive 2001/18 on the Deliberate Release into the Environment of Genetically Modified Organisms and Repealing Council Directive 90/220/EEC [2001] OJ L 106/1.

11 Meaning 'any intentional introduction into the environment of a GMO or a combination of GMOs for which no specific containment measures are used to limit their contact with and to provide a high level of safety for the general population and the environment', Article 2(3). A separate process applies to deliberate releases for purposes other than marketing (experimental releases) under Part B of the Deliberate Release Directive. Authorisation is granted by the single Member State (other Member

Modified Food and Feed (the 'Food and Feed Regulation'), which applies an obligation to seek authorisation for any GMO destined for use in food or feed.[12]

The regulatory framework thus varies in its detail according to whether we are concerned with a GMO used in food or (animal) feed, or a non-food GMO (such as cotton or flowers), although the two key pieces of legislation overlap. The Deliberate Release Directive applies to the 'placing on the market' of GMOs 'as or in products'.[13] The Food and Feed Regulation applies to 'GMOs for food use', 'food containing or consisting of GMOs' and 'food produced from or containing ingredients produced from GMOs', and so a broad range of agricultural and food products, including for example, tomato seeds and plants, tomatoes themselves, and tomato ketchup.[14] Many GMOs for placing on the market as or in products (Deliberate Release Directive) are also 'GMOs for food use' or 'food containing or consisting of GMOs' (Food and Feed Regulation). These GMOs may be authorised through a single application under the Food and Feed Regulation, but subject to environmental risk assessment under the Deliberate Release Directive. Alternatively, separate applications may be made under both pieces of legislation. 'Food produced from or containing ingredients produced from GMOs' is processed food that no longer contains actual GMOs, and so is not covered by the Deliberate Release Directive, but by the Food and Feed Regulation alone.

GMOs, such as flowers or cotton, that have no food or feed use are authorised under the Deliberate Release Directive alone. The applicant notifies the competent authority of the Member State where the GMO is to be placed on the market for the first time.[15] The notification must contain a range of information, including the environmental risk assessment carried out by the applicant, and a plan for monitoring the GMO following its release into the environment.[16] The notification is forwarded by the competent authority to the Commission and thence to the competent authorities of the other Member

---

States are informed) and release is limited to that Member State – experimental authorisation does not imply free movement. The Deliberate Release Directive also imposes obligations on labelling and traceability, discussed in Chapter 4 below.

12 Regulation 1829/2003 on Genetically Modified Food and Feed [2003] OJ L 268/1. For discussion of the new regimes, see Sara Poli, 'The Overhaul of the European Legislation on GMOs, Genetically Modified Food and Feed: Mission Accomplished. What Now?' (2004) 11 *Maastricht Journal of European and Comparative Law* 13; Maria Lee, *EU Environmental Law: Challenges, Change and Decision-Making* (Hart Publishing, 2005), Chapter 9. References will be made to food throughout this chapter; Chapter III of the Regulation applies similar provisions to feed.

13 Directive 2001/18, above n. 10, Article 1.

14 Regulation 1829/2003, above n. 12, Article 3(1)(a), (b) and (c).

15 Directive 2001/18, above n. 10, Article 13(1).

16 Directive 2001/18, above n. 10, Article 13.

States. This provision of information happens earlier in the process than it did under the 1990 Directive, which should provide an opportunity for discussion before positions are entrenched.[17] The initial competent authority examines the notification for compliance with the Directive, and prepares an assessment report, which is also circulated.[18] The assessment report indicates whether the GMO should (and under what conditions) or should not be placed on the market.[19] In the latter case, the application is rejected, with reasons,[20] but the applicant can make the same application to any other competent authority. In the former case, in the absence of a reasoned objection from another Member State or the Commission, the competent authority gives its consent, valid throughout the EU.[21]

The unwillingness of Member States to accept each other's risk assessments at this stage was a significant barrier to a peaceful authorisation process under the 1990 legislation, and continues to be so. Under the 2001 Deliberate Release Directive, a period is built into the legislation for Member States and the Commission to 'discuss any outstanding issues with the aim of arriving at an agreement'.[22] In the event of continued disagreement, however, we turn from mutual recognition of national risk assessments, discussed in the last paragraph, to a 'Community procedure'. Under this procedure, the Commission, following consultation of the European Food Safety Authority (EFSA), puts a draft decision to a 'regulatory committee' and, failing agreement in Committee, to Council.[23] If this leads to a decision that authorisation is appropriate, the initial competent authority must give its consent; in the absence of new information, it has no further discretion at this point.[24] This 'comitology' procedure is also the final stage of the decision making process under the Food and Feed Regulation, and so will be discussed further below, after examining first how 'risk assessment' proceeds under that Regulation.

The Food and Feed Regulation reduces the role of the national competent authorities in risk assessment compared with the Deliberate Release Directive

---

17 Under the 1990 legislation, above n. 5, Member States were just informed of the original Member State's decision on the application, Articles 12 and 13. See Estelle Brosset, 'The Prior Authorisation Procedure Adopted for the Deliberate Release into the Environment of Genetically Modified Organisms: The Complexities of Balancing Community and National Competences' (2004) 10 *European Law Journal* 555.

18 Directive 2001/18, above n. 10, Article 14(1) and (2).

19 Directive 2001/18, above n. 10, Article 14(3).

20 Directive 2001/18, above n. 10, Article 15(2).

21 Directive 2001/18, above n. 10, Article 15(3).

22 Directive 2001/18, above n. 10, Article 15(1).

23 Decision 1999/468 above n. 8, Article 5.

24 Case C-6/99 *Greenpeace v Ministère de l'Agriculture et de la Pêche* [2000] ECR I-1651 on similar wording in the earlier legislation.

and its predecessor. The initial application is sent to a national competent authority, but that authority simply passes the application to EFSA. EFSA, 'an independent scientific point of reference in risk assessment',[25] was (as discussed in Chapter 1) set up during the moratorium in response to a string of food scandals. Risk assessment is its primary tool and responsibility, and risk management is in principle for the political institutions.[26] EFSA's constitution emphasises scientific excellence and the independence of this science from political and industry influence. EFSA provided a useful source of (scientific) authority during the renegotiation of the GMO legislation. Whilst agencies at national level commonly 'borrow' the political legitimacy of the national political institutions, here the Commission is hoping to gain some legitimacy from the exercise of expertise by EFSA.

EFSA makes the application available to the Commission and other Member States, and the summary available to the public. EFSA has to provide an opinion on the application. In order to draw up this opinion, it *may* ask a national competent body to carry out a food safety assessment in accordance with the Food and Feed Regulation or an environmental risk assessment in accordance with the Deliberate Release Directive.[27] In the case of an overlap with the Deliberate Release Directive,[28] an environmental risk assessment as in that Directive is a mandatory part of the application, and EFSA is obliged to *consult* all national competent authorities.[29] If the application is for authorisation of 'seeds or other plant propagating material', EFSA '*shall* ask a national competent authority to *carry out*' the environmental risk assessment.[30] In most cases, the involvement of the national competent authorities in risk assessment is either at EFSA's discretion or simple consultation. Members of the different bodies of EFSA are not appointed by the Member States, and do not provide for representation of national interests, making this a firmly central body. Moreover, EFSA has only subsidiary concern for the environment,[31] and yet in many environmentally significant cases national

25 Regulation 178/2002 Laying Down the General Principles and Requirements of Food Law, Establishing the European Food Safety Authority and Laying Down Procedures in Matters of Food Safety [2002] OJ L 31/1, Recital 34.

26 The Regulation formalises the division of 'risk analysis' into separate stages of 'risk assessment', 'risk management' and 'risk communication' (Chapter 2, pp. 41–2). EFSA does have responsibilities in risk communication.

27 Regulation 1829/2003, above n. 12, Article 6(3)(b) and (c).

28 GMOs, or food containing or consisting of GMOs.

29 Regulation 1829/2003, above n. 12, Article 6(4).

30 Regulation 1829/2003, above n. 12, Article 6(3)(c), emphasis added.

31 Its purpose being to 'contribute to a high level of protection of human life and health, and in this respect take account of animal health and welfare, plant health and the environment', Regulation 178/2002, above n. 25, Article 22(3).

environmental authorities are only consulted. Nevertheless, the mandatory consultation undoubtedly provides at least an opportunity for the incorporation of national perspectives on risk assessment, as well as an important environmental perspective, although it does not look as if national competent authorities are falling over themselves to get involved.[32]

Centralisation of the 'scientific' stage of risk regulation is however subject to complex interactions between 'central' and 'national' experts. EFSA is required by its constitution to promote 'the European networking of organisations' operating in its area 'to facilitate a scientific cooperation framework by the coordination of activities, the exchange of information, the development and implementation of joint projects, the exchange of expertise and best practices'.[33] 'Networking' of risk assessors through agency structures invites national risk regulators and national perspectives on risk into the EU system. This attempts to compromise between the easier and more coherent decision making that goes with increased centralisation and the danger that too much centralisation damages the legitimacy of decisions.[34] The effectiveness of this compromise is clearly crucial in the particular political context of GMOs.

Networks do not resolve any problems related to the effectiveness or popular legitimacy of centralised bodies, particularly if the network is simply composed of like-minded elites. But when there is disagreement between national and central risk assessors, networks can at least address that disagreement. EFSA is required to 'exercise vigilance' in respect of 'any potential source of divergence' between its scientific opinions and those of other bodies.[35] There is an obligation to contact the body in question to ensure the sharing of scientific information in order to identify 'potentially contentious scientific issues'. There is then an obligation to cooperate in order either to resolve the divergence or to prepare and publish a joint document 'clarifying the contentious scientific issues and identifying the relevant uncertainties in the data'. The degree of pressure towards consensus will determine whether this process buries or exposes scien-

---

32 'EFSA experiences difficulties to find [competent authorities] willing to carry out the initial risk assessment', European Commission, *Report on the Implementation of Regulation 1829/2003 on Genetically Modified Food and Feed* COM (2006) 626 final, p. 10.

33 Regulation 178/2002, above n. 25, Article 36. This applies generally, so whether EFSA's opinion is issued under Regulation 1829/2003, above n. 12, or Directive 2001/18, above n. 10.

34 Giandomenico Majone, 'The Credibility Crisis of Community Regulation' (2000) 38 *Journal of Common Market Studies* 273, is of the view that centralisation would compound the 'credibility crisis'. See also Renaud Dehousse, 'Regulation by Networks in the European Community: The Role of European Agencies' (1997) 4 *Journal of European Public Policy* 246.

35 Regulation 178/2002, above n. 25, Article 30.

tific disagreement. But the opportunity for consensus is enhanced, and remaining disagreement is brought into the open. If the diverging opinion comes from a Community body, the joint document is 'presented to the Commission'; if it is from a national body, the document is simply prepared and made public. EFSA does not arbitrate between different approaches, and there is no hierarchy of scientific information in the legislation, for example according to whether it is nationally or centrally produced. But persistent disagreement, and associated uncertainty, is out in the open for political debate.

EFSA's general obligations of transparency on divergent opinions are reinforced by the obligation in the Food and Feed Regulation to state the reasons for its opinion, including the information on which the opinion is based, which in turn includes the responses of consulted competent authorities.[36] The Commission is clearly anxious to improve the liaison between EFSA and national competent authorities, 'inviting', in its report on the operation of the Food and Feed Regulation, EFSA to 'liaise more fully with national scientific bodies, with a view to resolving possible diverging scientific opinions with Member States'. This places hope in the informal techniques of governance that have been productive in less contentious areas. But, where the 'divergence' is not capable of resolution, EFSA is to 'provide more detailed justification, in its opinions on individual applications, for not accepting scientific objections raised by the national competent authorities'.[37] This demand by the Commission for greater justification of scientific opinions reflects considerable awareness of the contentiousness of its own decisions (based on EFSA opinions, and to date following them very closely), and arguably also awareness of how elusive uncontroversial facts are in this area.

EFSA's opinion is sent to the Commission and the Member States for political decision making. The publication of the opinion, with an opportunity for the public to 'make comments',[38] provides a further opportunity for scientific knowledge to be challenged, beyond the elite national regulatory community. The dichotomy between scientific risk assessment and political risk management is not absolute even on the face of the legislation, which provides for the opinion to contain conditions or restrictions on authorisation, including monitoring obligations,[39] a completeness that may increase the sense within the Commission that it is constrained by the EFSA decision.

---

36 Regulation 1829/2003, above n. 12, Article 6(6).

37 COM (2006) 626 final, above n. 32, p. 11. For detailed consideration of EFSA Opinions and their handling of disagreement, see Damian Chalmers, 'Risk, Anxiety and the European Mediation of the Politics of Life' (2005) 30 *European Law Review* 649.

38 Regulation 1829/2003, above n. 12, Article 6(7).

39 Regulation 1829/2003, above n. 12, Article 6(5)(e).

This marks the beginning of the comitology stage of decision making, which also applies during the 'Community procedure' of the Deliberate Release Directive (above p. 66). Comitology is formally a mechanism through which the Member States can supervise the Commission's exercise of implementing powers, although GMOs illustrate as well as any policy area how difficult it is to draw lines between 'mere' implementation and more politically sensitive legislation. The Commission submits a draft of measures to a regulatory Committee, consisting of Member State representatives and chaired by the Commission.[40] The draft has to take 'into account the opinion of the Authority [EFSA], any relevant provisions of Community law and other legitimate factors relevant to the matter under consideration'.[41] The Committee delivers its decision by qualified majority, and if the Committee gives a positive opinion the Commission adopts the decision. If not, the decision goes to Council. In the vast majority of cases in EU law, comitology committees agree with the Commission, and so the decision never reaches the Council. GMOs are, however, an anomaly in this respect, and that is no bad thing. Council consideration slows the process down, but at least ensures that authorisation of GMOs receives high-level political consideration. Because, although the committees are in principle political bodies and make political decisions, there is no necessary connection with public concerns. Comitology committees consist mainly of national technical experts. Sharing a common professional background and largely sheltered from public scrutiny, there is a danger that committees will reconceptualise political decisions as purely technical questions, ignoring their important political implications.[42] When the Committee does not agree with the Commission, the proposal is passed to the Council, where a decision is again taken on a qualified majority basis. If the Council fails to act (including a failure to reach a qualified majority decision in either direction) the Commission 'shall' adopt its decision, wording that suggests an absence of discretion.[43]

The authorisation that ended up the subject of the *Greenpeace*[44] litigation brought out the difficulties of comitology. A new procedure was introduced in

---

40 Regulation 1829/2003, above n. 12, Articles 7(1) and 35; Decision 1999/468 above n. 8, Article 5.

41 Regulation 1829/2003, above n. 12, Article 7(1).

42 See for example the contributions to Christian Joerges and Ellen Vos (eds), *EU Committees: Social Regulation, Law and Politics* (Hart Publishing, 1999). See also Decision 2006/512 above n. 8. Note that the European Parliament Committee on Environment, Public Health and Food Safety would have applied the regulatory procedure with scrutiny to authorisation of GM food; see Report 14 July 2007. This was not picked up in the text adopted by the European Parliament, 29 November 2007.

43 Decision 1999/468, above n. 8, Article 5.

44 Above n. 24.

1999,[45] but, as things stood before the moratorium, the Commission's draft could only be rejected by *unanimity* in Council.[46] And yet, the very fact that the comitology procedure had been brought into play meant that unanimity in Council was unlikely: at least one Member State had thought authorisation appropriate, and at least one had objected.[47] The Commission in the *Greenpeace* case was able to adopt its decision in spite of angry objections from a number of Member States, a European Parliament resolution against authorisation, and the positive approval of only one Member State in Council.[48] The ability of Council to reject the Commission's proposal by qualified majority under the current legislation obviously removes this extreme scenario. However, Member State disagreement has so far meant that the Council has been unable to reach a qualified majority either for or against the Commission's proposal.[49] This leaves considerable power in the hands of the Commission. The delicacy of the Commission's strong position in Comitology has long been recognised in a Commission declaration that in 'particularly sensitive sectors' the Commission will 'avoid going against any predominant position which might emerge within the Council'.[50] The very fact of an inability to reach qualified majority in Council, however, implies both that we have a 'sensitive sector', and that the 'predominant position' may not be obvious. Conflict between the Member States makes the default position crucial: national disagreement reverts the decision to the centre (the Commission), rather than back to the national level.

---

45 Decision 1999/468, above n. 8.

46 Council Decision 1987/373 [1987] OJ L 197/33.

47 See Tamara Hervey, 'Regulation of Genetically Modified Products in a Multi-Level System of Governance: Science or Citizens?' (2001) 10 *Review of European Community and International Environmental Law* 321. The only situation, other than a national change of view, in which unanimity would be possible is if the initial objection came from the Commission, not a Member State.

48 Commission Decision 1997/98 [1997] OJ L 31/69. See Hervey, above n. 47; Kieran St Clair Bradley, 'Institutional Aspects of Comitology: Scenes From the Cutting Room Floor' in Joerges and Vos, above n. 42.

49 The institutional framework within which decisions are taken continues to evolve. As well as the new regulatory procedure with scrutiny (above n. 42), the Lisbon Treaty amending the Treaty of European Union and the Treaty establishing the European Community ([2007] OJ C 306/01) introduces 'double majority voting' to replace the current calculations of qualified majority voting. If, as intended, this makes majorities easier to find, authority may eventually be returned to Council – albeit by an overwhelming of minorities rather than collaboration or compromise.

50 Declaration No. 3 on Council Decision 1999/468 Laying Down the Procedures for the Exercise of Implementing Powers Conferred on the Commission [1999] OJ C 203/1.

MAKING DECISIONS ON GMOs

The basic decision-making procedure for the authorisation of GMOs is outlined in the previous section. In this section, the intention is to study that procedure in a little more detail, along the themes of science, politics, internal market and finally how ideas of multi-level governance resonate in this area. It is, of course, artificial to divide the decision-making process along these lines, because it is far more fluid than that, and in particular it would be misleading to suggest that the science can be divided from the politics in any neat way. This approach does, however, to a large degree reflect the presentation of decision making in the legislation.

Scientific Risk Assessment

The two declarations that prompted the suspension of authorisations and the beginning of the moratorium refer to 'the need to put in place a tighter, more transparent framework, in particular for risk assessment, having regard to the specifics of European ecosystems'.[51] Disagreement between Member States under the 1990 Directive had meant that the more complicated and time-consuming 'Community' procedure had become the norm for decision making. A combination of detailed provisions on the environmental risk assessment under the Deliberate Release Directive and passing the decision on risk assessment to EFSA under the Food and Feed Regulation is one response to this refusal of Member States to accept each other's risk assessments.[52] Whilst this response has not yet been successful, the tightening-up of risk assessment in the new legislation is worth some examination.

Before the introduction of the new legislation, the most politically contested issue in respect of food had been the notion of 'substantial equivalence'. The 1997 Novel Foods Regulation provides an authorisation procedure for foods and ingredients that 'have not hitherto been used for human

---

51 Above n. 6.

52 Much less detail is provided on risk assessment in the Food and Feed Regulation, although the food safety assessment can be carried out by a national competent authority, Regulation 1829/2003, above n. 12, Article 6(3)(b), in accordance with Regulation 178/2002, above n. 25, Article 36. Regulation 178/2002 defines risk assessment as 'a scientifically based process consisting of four steps: hazard identification, hazard characterisation, exposure assessment and risk characterisation', Article 3(11). Risk assessment 'shall be based on the available scientific evidence and undertaken in an independent, objective and transparent manner', Article 6(2). More specifically, see EFSA, *Guidance Document of the Scientific Panel on Genetically Modified Organisms for the Risk Assessment of Genetically Modified Plants and Derived Food and Feed* (EFSA, 2006).

consumption to a significant degree within the Community', and that used to include GM food.[53] A 'simplified procedure' applied to foods derived from GMOs that were 'substantially equivalent' to existing foods in their 'composition, nutritional value, metabolism, intended use and the level of undesirable substances contained therein'.[54] The simplified procedure required only notification of the food by the applicant, and even during the 'moratorium' allowed for the authorisation of some GM foods.[55] The idea of substantial equivalence is highly controversial, primarily because it sidesteps the need for a scientific risk assessment.[56] The 'simplified procedure' has now been abandoned in respect of GM foods, on the basis that 'Whilst substantial equivalence is a key step in the procedure for assessment of the safety of genetically modified foods, it is not a safety assessment in itself'.[57] GM food now requires authorisation under the Food and Feed Regulation, which itself rests on risk assessment.

More generally, quite what a risk assessment involved under the 1990 legislation was disputed. The appropriate methodology for environmental risk assessment is spelled out in some detail in the Deliberate Release Directive.[58] Its scope is broad: environmental risk assessment means 'the evaluation of risks to human health and the environment, whether direct or indirect, immediate or delayed, which the deliberate release or the placing on the market of GMOs may pose', extending also to 'cumulative long-term effects', which refer to 'the accumulated effects of consents on human health and the environment, including inter alia flora and fauna, soil fertility, soil degradation of organic material, the feed/food chain, biological diversity, animal health and resistance problems in relation to antibiotics'.[59]

The coverage of the original legislation, that is, the grounds on which the GMO was to be assessed, was uncertain and often narrowly interpreted. A

---

53 Regulation 258/1997 concerning novel foods and novel food ingredients [1997] OJ L 043/1, Article 1(2).

54 Article 3(4).

55 See the discussion of early applications in Sarah Lieberman and Tim Gray, 'The So-called "Moratorium" on the Licensing of New Genetically Modified (GM) Products by the European Union 1998–2004: A Study in Ambiguity' (2006) 15 *Environmental Politics* 592, p. 605.

56 See discussion in Case C-236/01 *Monsanto Agricoltura Italia SpA v Presidenza del Consiglio dei Ministri* [2003] ECR I-8105. For a discussion of the ways in which concerns about substantial equivalence were raised, see Les Levidow, Joseph Murphy and Susan Carr, 'Recasting "Substantial Equivalence": Transatlantic Governance of GM Food' (2007) 32 *Science, Technology and Human Values* 26.

57 Regulation 1829/2003, above n. 12, Recital 6.

58 Directive 2001/18, above n. 10, Annex II.

59 Directive 2001/18, above n. 10, Article 2(8) and Annex II.

non-exhaustive list of potential adverse effects is provided in the new Deliberate Release Directive, in an effort to resolve the ambiguity of earlier legislation.[60] For example, the relevance of the herbicide regime associated with the GMO was disputed under the 1990 legislation, and it was argued by some Member States that this is an economic rather than an environmental issue. Even the Commission initially considered that 'the possible development of resistance . . . in insects cannot be considered an adverse environmental effect, as existing agricultural means of controlling such resistant species of insects will still be available'.[61] 'Changes in use or management' are now explicitly included in the 'indirect effects' to be assessed in environmental risk assessment.[62] Nor was it clear whether antibiotic resistance had to be assessed, an issue highlighted by the *Greenpeace* case. Greenpeace challenged the authorisation by France of Bt176 maize because France had not considered any potential problems caused by the presence of an antibiotic resistance marker gene. By contrast, other Member States had considered antibiotic resistance to be so significant as to require resort to the Directive's safeguard clause.[63] As well as listing 'resistance problems in relation to antibiotics' as a 'cumulative long-term effect' to be assessed,[64] the 2001 legislation explicitly requires the assessment of antibiotic resistance: antibiotic resistance marker genes have to be 'taken into particular consideration when carrying out an environmental risk assessment . . . with a view to identifying and phasing out antibiotic resistance markers in GMOs which may have adverse effects on human health and the environment'.[65] The wording is a little ambiguous, and although the intention seems to be to phase out antibiotic resistance markers (indeed a deadline of the end of 2004 is imposed),[66] there seems also to be a requirement to establish adverse effects case by case.[67]

---

60 Directive 2001/18, above n. 10, Annex II, C.2, para. 1.

61 Commission Decision 1997/98/EC, cited in *Greenpeace*, above n. 24, para. 11. See also Joseph Murphy, Les Levidow and Susan Carr, 'Regulatory Standards for Environmental Risks: Understanding the US–EU Conflict over GM Crops' (2006) 36 *Social Studies of Science* 133.

62 Directive 2001/18, above n. 10, Annex II, preamble, and paras C.2.1 and D.2.9.

63 See Hervey, above n. 47.

64 Directive 2001/18, above n. 10, Annex II, preamble.

65 Directive 2001/18, above n. 10, Recital 22, Article 4(2), Annex II.

66 Directive 2001/18, above n. 10, Article 4(2). See also European Commission, Staff Working Document, *Second Report on the Experience of Member States with GMOs Placed on the Market under Directive 2001/18/EC on the Deliberate Release into the Environment of Genetically Modified Organisms* SEC (2007) 274.

67 So see European Commission, *Proposal for a Council Decision concerning the placing on the market in accordance with Directive 2001/18/EC of a potato product genetically modified for enhanced content of the amylopectin component of starch*

Another change is the explicit inclusion of the precautionary principle in the Deliberate Release Directive.[68] The precautionary principle is in any event an autonomous principle of Community law,[69] and had been applied in respect of the earlier legislation,[70] so this is arguably more politically than legally significant. The legislation, however, couples the precautionary principle with apparently stringent substantive obligations for authorisation.[71] The Deliberate Release Directive requires that, 'in accordance with the precautionary principle', Member States 'ensure that all appropriate measures are taken to avoid adverse effects on human health and the environment which might arise from the deliberate release or the placing on the market of GMOs'.[72] 'Appropriate' measures clearly leaves room for judgment, but Recital 47 provides that 'The competent authority should give its consent only after it has been satisfied that the release will be safe for human health and the environment', and the Treaty itself provides that these measures 'take as a base a high level of protection'.[73] This suggests quite a low tolerance of adverse effects, although as discussed in Chapter 2 (p. 46) seeking 'zero risk' in the sense of proof of safety is not generally permitted in EU law. Even without challenging this bar on the consideration of unknowns (and it is not obvious why unknowability ought *never* to be relevant[74]), and without doing any violence to the case law, seeking to reduce an *identified* (not hypothetical) risk to zero is both permissible,[75] and a possible interpretation of the Deliberate Release Directive.

COM (2007) 336 final, proposing authorisation of a GMO using an antibiotic resistance marker gene, relying on scientific advice as to safety. See COM (2004) 575 final on the development of 'clear and transparent criteria when assessing whether a specific [antibiotic resistance marker] is to be considered harmful', Annex 4.

68 Directive 2001/18, above n. 10, Recital 8, Article 1, Article 4. There is no explicit reference to the precautionary principle in the Food and Feed Regulation, but it is contained in the General Food Regulation, Regulation 178/2002, above n. 25, Article 7.

69 Stemming from and applying to all Treaty responsibilities for public health, safety and the environment; see for example Cases T-74/00, T-76/00 and T-141/00 *Artegodan and Others v Commission* [2002] ECR II-4945, para. 184.

70 See *Greenpeace*, above n. 24, and *Monsanto*, above n. 55.

71 Theofanis Christoforou, 'The Regulation of Genetically Modified Organisms in the European Union: The Interplay of Science, Law and Politics' (2004) 41 *Common Market Law Review* 637.

72 Directive 2001/18, above n. 10, Article 4.

73 Article 95(3), EC Treaty.

74 Note that Nicholas de Sadeleer, 'The Precautionary Principle in EC Health and Environmental Law' (2006) 12 *European Law Journal* 139, argues that the stricter approach applies to the withdrawal of products only, p. 163.

75 Case C-121/00 *Hahn* [2002] ECR I-9193; Case C-286/02 *Bellio v Prefettura di Treviso* [2004] ECR I-3465.

The Food and Feed Regulation contains even stronger language than the Directive: a GMO must not 'have adverse effects on human health, animal health or the environment'; 'mislead the consumer'; or 'differ from the food which it is intended to replace to such an extent that its normal consumption would be nutritionally disadvantageous for the consumer'.[76] As with the Deliberate Release Directive, a requirement to establish 'no adverse effects' presumably does not require proof by an applicant of absolute safety. The initial proposal for the Food and Feed Regulation had provided that GM food must not 'present a risk for human health or the environment'.[77] The changed wording in the final version more clearly focuses on harm rather than risk. It might nevertheless be argued that positive evidence of *any* adverse effect would preclude authorisation – a mandatory 'zero tolerance' of harm. Combined with the precautionary principle, particularly if the precautionary principle develops in such a way as to *require*, as well as protect, precautionary action,[78] the prohibition on adverse effects could provide a very serious legal limitation on the authorisation of GMOs. However, all human activity affects the environment, and much agricultural activity could be said to have an adverse effect at some level. The intention is unlikely to be to design legislation in a way that would require applications for authorisation to be rejected for most, if not all, of the current generation of GMOs, and proportionality is another general principle of EU law. 'No adverse effects' seems to imply comparison with conventional crops. Certainly under the Deliberate Release Directive, the characteristics of a GMO that has potential adverse effects 'should be compared to those presented by the non-modified organism from which it is derived and its use under corresponding situations'.[79] And EFSA Guidance on risk assessment provides for comparison with 'non-GM counterparts': the 'underlying assumption' is that 'traditionally cultivated crops have gained a history of safe use for the normal consumer or animal and the environment'.[80] Any such comparison involves

---

76 Regulation 1829/2003, above n. 12, Article 4(1).

77 See European Commission, *Proposal for a Regulation of the European Parliament and of the Council on Genetically Modified Food and Feed* COM (2001) 425 final, Article 4(1).

78 Case C-127/02 *Landelijke Vereniging tot Behoud van de Waddenzee v Staatssecretaris van Landbouw, Natuurbeheer en Visserij* [2004] ECR I-7405 on national implementation of EC law; in respect of EC action, see Cases T-74/00, T-76/00 and T-141/00 *Artegodan and Others v Commission* [2002] ECR II-4945. See Chapter 2, pp. 43–8.

79 Directive 2001/18, above n. 10, Annex II, para. B. Food is compared with its 'conventional counterpart' under Regulation 1829/2003, above n. 12, Article 5(3)(f), which is, however, concerned particularly with the 'consumer interests' conditions of authorisation, rather than safety.

80 EFSA, above n. 52, p. 12.

assumptions and uncertainties, not only in respect of the GMO, but also in respect of conventional agricultural practices.[81] At the extreme, for example, comparison with organic agriculture is likely to produce a very different conceptualisation of 'adversity' from comparison with highly intensive farming. Given that some Member States have a greater commitment to organic farming than others, this is a difficult normative question. But in short, although the level of acceptable safety/risk is open, the language of the legislation is strong.

The limitation of authorisation to a ten-year period (with the possibility of renewal) is another significant element of risk regulation, meaning that individual decisions will always be revisited.[82] This importantly acknowledges the changing scientific basis for understanding GMOs, and the need to learn from experience in this area. The limited authorisation is reinforced by the obligation on the authorisation holder to monitor GMOs after their release,[83] for potential adverse effects identified in the risk assessment and for unanticipated effects, and to inform decision makers of new information that might influence the safety evaluation.[84] Implicitly, these provisions acknowledge the uncertain basis of the science on which the decision is made, providing for the renewal of knowledge and learning through the regulatory process, and allowing assumptions to be checked following experience.

These are all undoubted improvements from the perspective of those concerned about the risks of GMOs, balanced for industry by promises, not yet realised, about improved predictability and efficiency, as well as improved consumer confidence. The risk assessment process is of course not without its critics. In particular, the very idea of case-by-case analysis has its limitations. The legislation refers explicitly to 'cumulative' effects, but assessment of cumulative effects on a case-by-case basis will be very complex as the use of GMOs expands, and as gene stacking (in organisms with many modifications) increases. Case-by-case risk assessment also assumes that the observation of changes in any single organism, or even in

---

81 See the discussion of the UK's farm-scale evaluations in Chapter 2, pp. 51–3.

82 Directive 2001/18, above n. 10, Article 15; Regulation 1829/2003, above n. 12, Article 7.

83 Monitoring is an automatic condition of authorisations covered by Directive 2001/18, above n. 10, Article 20; under Regulation 1829/2003, above n. 12, monitoring may be one of the conditions of authorisation. Detail is provided in Council Decision 2002/811/EC establishing guidance notes supplementing Annex VII to Directive 2001/18/EC on the deliberate release into the environment of genetically modified organisms [2002] OJ L 280/27.

84 Regulation 1829/2003, above n. 12, Article 9(3); Directive 2001/18, above n. 10, Article 20.

the relationship between that organism and other organisms in its proposed environment, reveals everything that regulators need to know. The impact of many organisms, with many modifications, on many ecosystems is highly complex and not likely to be fully addressed in advance of release. This is largely the problem of 'indeterminacy' discussed in Chapter 2 (p. 30), and the real-world impact of organisms in ecosystems is enormously complex. Similarly, social systems are complex and relevant. Laboratory tests might suggest that a particular level of pesticide/herbicide application is appropriate, but the actual behaviour of farmers is likely to depend on a range of economic, environmental and social factors, including liability and enforcement. And of course, as discussed above (p. 75) 'ignorance', the many things we do not and cannot know about GMOs, would probably be classed as a 'hypothetical' risk, not, according to the courts, a suitable basis for regulation in any event.

These questions of indeterminacy and ignorance are social questions that can be addressed by the political decision makers as readily as the scientists – but only if the risk assessment is not deemed to have provided all the relevant information on the facts. The danger is that the political decision makers will consider themselves constrained by scientific advice, and that the gaps and value judgments inherent in that scientific advice will be rendered invisible by the focus on science.

The information required for the risk assessment under both pieces of legislation is provided by the applicant, following the detailed requirements in the legislation and guidance. The placing of responsibility on the applicant has certain advantages, requiring an investigation of 'direct or indirect, immediate or delayed' effects of GMOs on the environment at the expense of the party seeking to profit from the GMOs, possibly enhancing awareness of risk within that organisation, which is best able to address the risk. The concern that the applicant thus gets to set the agenda for the risk assessment is almost too obvious to state. Although ameliorated by the detail as to how the process should be carried out, and the multiple opportunities for scrutiny, this emphasises the inequality of knowledge and power between regulator and industry.[85] There is also considerable scope for the applicant to seek to withhold information as commercially confidential, and some interest groups have expressed concern about the sole responsibility of the consent holder for monitoring.[86] To the

---

85 See Neil Gunningham, 'Regulating Biotechnology: Lessons from Environmental Policy' in Han Somsen (ed.), *The Regulatory Challenge of Biotechnology: Human Genetics, Food and Patents* (Edward Elgar, 2007).

86 European Commission, *Second Report on the Experience of Member States with GMOs Placed on the Market under Directive 2001/18/EC on the Deliberate Release into the Environment of Genetically Modified Organisms* COM (2007) 81 final, p. 6. See Javier Lezaun, 'Creating a New Object of Government: Making

extent that concern about GMOs is mistrust of the industry's motives as well as its products, risk assessment on this basis is always likely to fall short.

And, finally, a mechanism for requiring the withdrawal of unauthorised GMOs that come onto the market is conspicuous by its absence.[87] In 2006, shipments of rice from the US were contaminated by GM rice that had not been authorised in the EU. There is no threshold level at which the presence of unauthorised GMOs in non-GM or authorised products is acceptable,[88] and the Commission issued emergency decisions under Article 53 of the General Food Regulation ordering the withdrawal of the unauthorised rice.[89] Article 53, however, applies where food 'is likely to constitute a serious risk to human health, animal health or the environment'. Whilst this was the best option available to the Commission, it was something of a stretch to argue, in the absence of scientific evidence, that the rice constituted a 'serious risk'. Taking any action at all against this GM rice required an assumption of harm in the absence of authorisation. This is indicative of the fixation of the legislation with a certain type of risk. Withdrawing unauthorised GMOs may not be about the specific risk posed by the individual GMO, but about the integrity of the legislation, and more holistic questions of the risk posed by ineffective regulation of GMOs.

To summarise, the risk assessment process has been very considerably tightened up in the legislation, responding to critical discussion of the regulatory optimism of the 1990s.[90] It is thorough, wide ranging, and open to competing explanations. There are, however, lingering concerns about the adequacy of the risk assessment paradigm as the main aspect of decision making. This is to some extent mitigated by the explicit provision of a space for political decision making, discussed in the next section.

Genetically Modified Organisms Traceable' (2006) 36 *Social Studies of Science* 499, pp. 516 and onwards, for a discussion of how commercial confidentiality limits regulators.

87 Article 45 of Regulation 1829/2003, above n. 12, requires the Member States to lay down penalties for breach, 'and to take all measures necessary to ensure they are implemented'.

88 Although there was a transitional provision in respect of food and feed, Regulation 1829/2003, above n. 12, Article 47. A transitional period is also provided in respect of GMOs withdrawn from the market.

89 Commission Decision 2006/578 on emergency measures regarding the non-authorised genetically modified organism LL RICE 601 in rice products [2006] OJ l 230/8, updated by Commission Decision 2006/601 [2006] OJ L 244/27. Action has also been taken on Bt10 maize and papaya; see Commission, above n. 37.

90 See Murphy, Levidow and Carr, above n. 61.

Beyond Science?

The authorisation of GMOs is deeply political, and the final part of the decision making process involves both the Commission and the Member States, through comitology as discussed above, and is an explicitly political decision. The first and very important question has to be the basis on which they can legitimately take that political decision.

As in the UK, discussed in Chapter 2 (pp. 49–59), EU policy has developed to a point where it acknowledges the relevance of non-safety concerns in debate about GMOs. The *Life Sciences Strategy* acknowledges the breadth of concern on these new technologies, and provides that 'new ethical or societal implications' should be 'addressed proactively and with a broad perspective'.[91] EU policy clearly recognises that 'societal scrutiny and dialogue should accompany and guide the development of life sciences and biotechnology'.[92] Whilst confirming and even emphasising the importance of dialogue, however, the *Life Sciences Strategy* seems to demonstrate a frustration that the economic potential of the life sciences is not being achieved because of public mistrust. There is a resulting ambivalence to the value of public participation: 'We shall also strive for a balanced and rational approach, distinguishing between real issues, on which we must act, and false claims'.[93] The Commission seems to think that distinguishing between 'real issues' and 'false claims' will be unproblematic. Space for concerns not backed up by science, including concerns about scientific ignorance, is probably limited. The Commission elsewhere refers to the need for 'a systematic impact analysis on the benefits and risks of biotechnology in order to support a structured and evidence based societal dialogue and policy making process',[94] rather missing the point that it is precisely the profound uncertainty about these issues that intensifies concern.

Hinting at a control that would exclude much of the rather chaotic protest against GMOs, the Strategy says that 'Dialogue in our democratic societies should be inclusive, comprehensive, well informed and structured'.[95] Moreover, a considerable amount of the discussion of openness and accountability is related to the 'general need to enhance public trust in the role of

---

91 Above n. 1, p. 20.

92 *Life Sciences Strategy,* above n. 1, p. 17 and onwards, and generally section 4.

93 *Life Sciences Strategy,* above n. 1, p. 18.

94 European Commission, Staff Working Document, *Communication on the Mid-term Review of the Strategy on Life Sciences and Biotechnology* SEC (2007) 441, p. 33.

95 *Life Sciences Strategy,* above n. 1, p. 18.

science in our societies' and to increase confidence in the regulation. This model of persuasion, rather than participation, is incapable of capturing the nuances of concern about GMOs. And, although it suggests, picking up on Chapter 1, that the Commission acknowledges the reality of public concern, it provides no assurance that the Commission recognises the real substantive issues behind the political fact of public concern.[96]

Consistently with the policy, both pieces of legislation provide for the consultation of an ethical committee, and for public comments prior to authorisation. The legislation also just touches on socioeconomic issues, requiring the Commission, in its periodic reports on the legislation, to include a chapter on the socioeconomic advantages and disadvantages of each category of GMO, 'which will take due account of the interests of farmers and consumers'.[97]

The legislative provisions for consultation of an ethical committee (the European Group on Ethics in Science and New Technologies) must constitute an acceptance that agricultural biotechnology is not only about risk.[98] Quite what the impact of these consultations might be is not clear. The Deliberate Release Directive is explicit that this consultation does not affect the Directive's 'administrative procedures'.[99] The committee has not so far been active in the field of GMOs, but its existence does assume that individual modifications (modifications of animals seem most likely to provoke discussion) can be considered in terms other than 'risk'.[100] As well as providing for the consideration of ethical issues during EU decision making, the legislation confirms the 'competence' of Member States 'as regards ethical issues'.[101] Moreover, 'Member States may take into consideration ethical aspects when

---

96 See Brian Wynne, 'Creating Public Alienation: Expert Cultures of Risk and Ethics on GMOs' (2001) 10 *Science as Culture* 445, on the UK approach to GMOs.

97 Directive 2001/18, above n. 10, Recital 62 and Article 31(7). The Commission's first report on the Directive discussed the potential negative socioeconomic impact of regulation *on* the biotechnology industry (concern about regulatory burden and the possible impact on the EU research base), rather than any of the potential negative socioeconomic impacts brought about *by* the biotechnology industry, COM (2004) 250 final, above n. 2, Annex 4. It also discussed coexistence, on which see Chapter 4.

98 Directive 2001/18, above n. 10, Recital 57, Article 29; Regulation 1829/2003, above n. 12, Article 33.

99 Directive 2001/18, above n. 10, Article 29(3) – the same proviso applies to the consultation of scientific committees, Article 28.

100 The objectives of the Food and Feed Regulation embrace also 'animal welfare', Article 1. There is no specific legislation on GM animals, and nothing specific in the existing legislation on the subject.

101 Directive 2001/18, above n. 10, Article 29; Regulation 1829/2003, above n. 12, Recital 42.

GMOs are deliberately released or placed on the market as or in products',[102] and the Council has noted that 'the ethical acceptability of some areas of biotechnology is related to the diversity among Member States and is governed by national law in accordance with the principle of subsidiarity'.[103] It really is not clear how this is going to work. The implication seems to be that Member States may restrict the free movement of goods on ethical grounds, but it is difficult to see how this fits in with internal market law in anything other than possibly the most extreme cases. The attraction of centralised 'expertise' on this matter, and indeed a common European 'bioethics', is rather apparent.

It would be strange if the results of the legislative opportunities for the public to make comments to the Commission were not relevant to the final decision.[104] But, matching the ambivalence in the policy, the provision for public participation in the legislation is not particularly ambitious, limited to this opportunity to make comments. Although there is no limitation on the nature of the comments that can be made, any more ambitious approach to public involvement is left to the Member State. Feeding the results of national public involvement into the EU-level decision-making process will be difficult. The legislation concentrates primarily on access to information, although environmental interest groups have expressed concern that the 'commercial confidentiality' exception is leading to access being denied. In the case of the Deliberate Release Directive, even the access provisions are limited to the summary of the initial notification and *positive* (only) assessment reports, as well as the opinions of scientific committees consulted.[105] The Deliberate Release Directive does, however, provide for a public register of the location of GMOs.[106] The information compulsorily released under the Food and Feed Regulation includes the application for authorisation, the EFSA[107] and competent authority opinions, monitoring reports and information from the authorisation holder.[108] The format of the information and the means of publicity are left largely open: a summary of the applica-

102 Directive 2001/18, above n. 10, Recital 9.

103 *Council Conclusions and Roadmap*, above n. 1, p. 12.

104 Directive 2001/18, above n. 10, Article 24(1). Regulation 1829/2003, above n. 12, Article 6(7). Directive 2001/18, Recital 46 provides that comments of the public should be taken into consideration in drafts to the Regulatory Committee. See also Regulation 178/2002, above n. 25, which provides for 'open and transparent public consultation, directly or through representative bodies', Article 9.

105 Directive 2001/18, above n. 10, Article 24.

106 Directive 2001/18, above n. 10, Article 31(3)(b).

107 Regulation 1829/2003, above n. 12, Article 6(7).

108 Regulation 1829/2003, above n. 12, Article 29(1).

tion and a public register for the logging of authorisations are concessions to accessibility of information.[109]

The Food and Feed Regulation most explicitly moves beyond scientific risk assessment in its authorisation provisions. It recognises that 'in some cases, scientific risk assessment alone cannot provide all the information on which a risk management decision should be based'.[110] The Commission's draft decision can take account of not only the EFSA opinion and relevant provisions of Community law but also 'other legitimate factors relevant to the matter under consideration'.[111] Similar wording appears in the General Food Regulation, according to which 'other factors legitimate to the matter under consideration' include 'societal, economic, traditional, ethical and environmental factors and the feasibility of controls'.[112] With food generally, these factors will probably as often *permit* traditional food such as non-pasteurised cheese, potentially frowned upon by scientific risk assessment, as they will *restrict* products, which is more likely with GMOs. Most importantly, however, the Food and Feed Regulation on its face provides some real space for the incorporation into decisions of values and concerns that go beyond technical and scientific issues.

The purposes and objectives of GMOs, which as discussed in Chapters 1 and 2 are a central part of the public debate, might usefully fit into the 'other legitimate factors' rubric. The European Parliament had wanted the legislation to require applicants to provide 'justification of the social desirability of the objective of the proposed deliberate release and assessment of possible alternatives to attain the same objectives', but failed to convince the other legislators.[113] This sort of gap is common in environmental and public health regulation, reflecting a simple and usually unspoken assumption that growth or progress through technological development is necessary and desirable. The restriction of a new technology needs to be justified, not its introduction. And

109 Regulation 1829/2003, above n. 12, Article 28. Note also Commission Decision 2004/204 Laying Down Detailed Arrangements for the Operation of the Registers for Recording Information on Genetic Modifications in GMOs, Provided for in Directive 2001/18/EC [2004] OJ L 65/20.

110 Regulation 1829/2003, above n. 12, Recital 32.

111 Regulation 1829/2003, above n. 12, Article 6.

112 Regulation 178/2002, above n. 25, Recital 19. The possible content of other legitimate factors is not spelled out in Regulation 1829/2003, above n. 12. See Marsha Echols, 'Food Safety Regulation in the European Union and the United States: Different Cultures, Different Laws' (1998) 4 *Columbia Journal of European Law* 525, for a discussion of US and EU divergences on 'traditional foods'.

113 Gregory Shaffer and Mark Pollack, 'Regulating Between National Fears and Global Disciplines: Agricultural Biotechnology in the EU' (2004) *Jean Monnet Working Paper* 10/04.

without an assessment of the desirability of or need for the particular application of GMOs, the regulation cannot completely, or even partially, address the concerns outlined in Chapter 2. Nor would this understanding of 'other legitimate factors' be out of line with more general case law, although the situation is not entirely clear. In a case on Danish measures restricting the use of additives (sulphites, nitrites and nitrates) in food, the Court (by contrast with the Commission) accepted that 'technological need is closely related to the assessment of what is necessary in order to protect public health': 'In the absence of a technological need justifying the use of an additive, there is no reason to incur the potential health risk resulting from the authorisation of the use of that additive'.[114]

Lack of societal need for an innovation has also come up in cases on food enriched with vitamins or minerals, when a particular Member State considers that these vitamins or minerals are not deficient in the ordinary diet. Again, absence of such a need cannot on its own 'justify a total prohibition', but might be relevant along with information on risk.[115] Whilst the legislative context in these cases may provide greater support than 'other legitimate factors' for a role for need,[116] it is important that the case law supports the balancing of small or uncertain risks against small or uncertain benefits. The idea that purpose is relevant to risk resonates strongly with the public debate on GMOs. This line of case law, even if not used routinely, provides a useful marker in respect of GMOs.

And using the 'other legitimate factors' formula to take account of public views more generally is also not totally out of line with the case law. In *Fedesa*, famously, when faced with 'divergent appraisals by the national authorities', the Council acted within the limits of its discretion in banning the use of certain hormones in cattle, 'and respond[ing] in that way to the concerns expressed by the European Parliament, the Economic and Social Committee and by several consumer organizations'.[117] Where there are 'divergent appraisals' of risk, there can be no expectation 'that a prohibition . . . could be based on scientific data alone'.[118] So the public perceptions represented by the

---

[114] Case C-3/00 *Denmark v Commission* [2003] ECR I-2643, para. 82. This was a case under Article 95(4).

[115] Case C-24/00 *Commission v France* [2004] ECR I-1277, paras 59–60; Case C-192/01 *Commission v Denmark* [2003] ECR I-9693 para. 54; Case C-211/03 *HLH Warenvertriebs v Germany* [2005] ECR I-5141 states that the need for a development can 'play a role' in risk assessment, para. 69.

[116] In particular, note that a 'reasonable technological need' is a condition for the approval of food additives, *Denmark*, above n. 114, para. 6, whilst the benefits of biotechnology seem broadly to be assumed.

[117] Case 331/88 *R v MAFF ex parte Fedesa* [1990] ECR I-4023, para. 9.

[118] *Fedesa*, above n. 117, para. 10.

Parliament, Economic and Social Committee and consumer organisations were at least relevant in the final decision. Moreover, as Christopher Hilson points out, these 'divergent appraisals' do not seem to have been supported by scientific evidence, or at least none that was presented to the Court.[119] It is arguable that the increasing willingness of the Court of First Instance (CFI) and the European Court of Justice (ECJ) to police the scientific evidence used in decision making, and so the scientific background to the 'divergent appraisals', means that these elements of the decision will not survive more recent case law. Nevertheless, *Fedesa* continues to provide a powerful argument for political resolution in the face of disagreement based on scientific uncertainty. And even in more recent and scientifically focused decisions such as *Pfizer*, discussed in Chapter 2, the CFI states that 'the restoration of consumer confidence can . . . also be an important objective which may justify even substantial economic consequences for certain traders'.[120] In *Pfizer*, the CFI rejected the argument that the contested regulation 'was adopted with the sole aim of creating a favourable political impression in the press and with public opinion', holding that 'the contested regulation pursues, above all else, public health objectives'.[121] *Pfizer* is far narrower than *Fedesa*, both because it demands *sufficient* scientific information and because the consumer confidence objective is clearly considered secondary to public health protection. And other case law reminds us that consumer habits change, and that the internal market contributes to such change: national rules must not 'crystallise given consumer habits so as to consolidate an advantage acquired by national industries'.[122]

Decision makers do not apparently have to feign deafness to political considerations, but *Pfizer* and subsequent cases very strongly encourage political decision makers to support their decisions by reference to scientific risk assessment.[123] The CFI held that when a Community institution departs from the opinion of a scientific committee, it must provide 'specific reasons for its findings', and moreover that the reasons 'must be of a scientific level at least

119 Christopher Hilson, 'Beyond Rationality? Judicial Review and Public Concern in the EU and the WTO' (2005) 56 *Northern Ireland Legal Quarterly* 320 is a very useful discussion of the treatment of public concern.

120 Case T-13/99 *Pfizer Animal Health SA v Council* [2002] ECR II-3305, para. 462.

121 *Pfizer*, above n. 120, para. 462.

122 Case 170/78 *Commission v United Kingdom* [1980] ECR 417, cited in a number of cases.

123 Note that Elizabeth Fisher, *Risk: Regulation and Administrative Constitutionalism* (Hart Publishing, 2007), Chapter 6, identifies a change of approach following European Commission, *Communication on the Precautionary Principle* COM (2000) 1 final.

commensurate with that of the opinion in question'.[124] Science must be fought with more science. *Pfizer* deals with legislation that does not contain the 'other legitimate factors' formula, so it applies directly only to those GMOs assessed under the Deliberate Release Directive alone. On the face of the Food and Feed Regulation, if the Commission's draft decision differs from the EFSA opinion, the Commission simply has to 'provide an explanation for the difference', and presumably that 'explanation' can be based on other legitimate factors.[125] But, notwithstanding the wording of the Food and Feed Regulation, *Pfizer* and cases like it illustrate a risk philosophy that may tend to reduce the scope of other legitimate factors. The approach to safeguard clauses, discussed below (p. 89), reinforces the conclusion that science is the prime source of authority in the regulation of GMOs.

The case law is ambiguous, and it would be wrong to suggest that it demands only scientific justification of regulation. The legal context of the regulation, however, especially *Pfizer* and the line of case law following that decision, tends to marginalise decision-making criteria other than science. Basing a decision on other legitimate factors would be a brave step. A further very significant restriction on the potential breadth of other legitimate factors is the purpose of the legislation. Administrative powers must be exercised exclusively or primarily for purposes for which they were granted.[126] The objectives of the Deliberate Release Directive are the protection of human health and the environment.[127] There is ample opportunity for debate on safety, but, more problematically, safety seems to be the only or primary point of engagement. The Food and Feed Regulation embraces broader objectives than the Deliberate Release Directive, aiming to 'provide the basis for ensuring a high level of protection of human life and health, animal health and welfare, environment and consumer interests in relation to genetically modified food and feed, whilst ensuring the effective functioning of the internal market'.[128] This takes us beyond questions of safety for human and animal health and the environment. Consumer interests could conceivably include long-term and inchoate issues such as the gradual impact on consumers of the extension of monocultures, or increased reliance on a small number of patent holders. The conditions for authorisation found in the Regulation, however, are somewhat narrower: GMOs must not 'have adverse effects on human health, animal health or the environment', 'mislead the consumer' or 'differ from the food which it is intended to replace to such an extent that its normal

124 *Pfizer*, above n. 120, para. 199.
125 Regulation 1829/2003, above n. 12, Article 7.
126 See for example *Fedesa*, above n. 117.
127 As well as approximation of laws, Directive 2001/18, above n. 10, Article 1.
128 Regulation 1829/2003, above n. 12, Article 1(a).

consumption would be nutritionally disadvantageous for the consumer'.[129] Moreover, labelling seems to be the primary tool for consumer protection. For example, the Regulation specifies that the applicant must either establish in its application that the GM food is nutritionally no different from its conventional counterpart or suggest appropriate labelling.[130] Any 'ethical or religious concerns' are also explicitly linked with labelling.[131] It seems unlikely that the 'consumer interests' rationale will feed into the question of whether or not a GMO should be authorised very often, if at all.[132]

This return to questions of safety in the objectives of authorisation considerably restricts the potential for other legitimate factors to open up decision making. Nor does the precautionary principle provide much support for reliance on 'other legitimate factors'. Once we accept that science cannot provide authority for a decision based on incontrovertible 'facts', alternative sources of legitimacy, alternative reasons, must be sought. However, as discussed in Chapter 2, the EU approach to the precautionary principle, if anything, tends to confirm the likely privileging of scientific and technical information. A measure cannot be based on a 'purely hypothetical' approach to the risk, founded on a 'mere conjecture which has not been scientifically verified'.[133] Perhaps it is paradoxical, but the precautionary principle, whilst recognising the inherent limitations of science, relies on as complete a scientific risk assessment as possible, on 'the most reasonable scientific evidence and the most recent results of international research'.[134] The precautionary principle, as interpreted in EC law, provides no obvious refuge for the incorporation of ignorance or indeterminacy into the risk management process, let alone ethical or socioeconomic concerns. However, it should certainly be argued that what the Court in *Pfizer* dismissed as 'hypothetical' risk, a concern about ignorance, constitutes an 'other legitimate factor' under the Food and Feed Regulation. This is enormously important, raising the possibility for the real nature of uncertainty to be relevant to decision making. That is not to say

---

129 Regulation 1829/2003, above n. 12, Article 4(1).

130 Regulation 1829/2003, above n. 12, Article 5(3)(f).

131 Regulation 1829/2003, above n. 12, Article 5(3)(g).

132 See also Joanne Scott, 'European Regulation of GMOs and the WTO' (2003) 9 *Columbia Journal of European Law* 213. The Commission has suggested that future developments, for example a genetic modification that makes food look fresh when it is not, or makes a food look like a different product, could feed into the authorisation process rather than labelling, *Response from the Commission to Comments submitted by WTO Members under Either or Both G/TBT/N/EEC/6 and G/SPS/n/EEC/149*, 26 July 2002.

133 Case T-70/99 *Alpharma Inc v Council* [2002] ECR II-3495, para. 156; *Pfizer*, above n. 120, para. 143.

134 *Monsanto*, above n. 56.

that it would be decisive or even weighty. But simply bringing uncertainty into the balance against, for example, the uncertain benefits of GMOs, allows for a more realistic decision.

The 'other legitimate factors' formula is a hugely significant innovation. The legal framework within which it operates, however, leads to real ambivalence about its potential, and of course the influence of the WTO, discussed in Chapter 6, is important here. Even beyond legal emphasis on scientific rather than political decision making, relying on scientific advice rather than political judgment appeals to political decision makers for its appearance of neutrality and objectivity. This applies particularly strongly in the EU context, where conflicts and political disagreement are multiplied, and the appeal of an apparently disinterested arbiter is enormous. The distinction in the legislation between scientific assessment and political 'other legitimate factors' may reinforce the misperception of scientific conclusions as inevitable and value free, untainted by politics and 'other factors', legitimate or not. Relying on political decision making is of course difficult, and rife with potential for bias and abuse, not to mention more understandable error and inefficiency. However, it is a mistake to think that scientific information can escape these traps.

It is rather more difficult than it sounds to outline the permissible grounds for a decision on whether or not to authorise a GMO. In respect of food and feed at least (the 'other legitimate factors' formula only explicitly applies to food and feed), the grounds for a decision are potentially open-ended. The 'other legitimate factors' formula might in principle be relied on to bring a range of issues, from 'hypothetical' risk, to the purpose of the GMO, to corporate control of the food sector, into the regulatory fold. The legal and political context of decision making, however, is far more ambiguous, and the most recent case law provides considerable incentives for the explanation of decisions by reference to (perhaps minority) scientific evidence, rather than explicitly on the basis of political judgment. That ambiguity is enhanced by the Commission's adherence to EFSA opinions in the applications on which it has so far issued a draft Decision.

Nevertheless, the 'other legitimate factors' formula does provide an answer to those who complained about the narrow basis for decision making under the earlier legislation. The lingering ambiguity is an indication of how deeply entrenched the earlier, narrow approach is. Rather than a complete change to the technocratic nature of decision making, non-scientific considerations have been added to existing frameworks. Legislation that emphasises the importance of robust science, of transparent reason giving and of good evidence is, of course, no bad thing. But if the 'other legitimate factors' formula prospers, a politically reasoned decision could provide an accountable and sophisticated response to the reality of GM technology. This applies whether the decision grants or withholds authorisation, and it would be naïve not to expect such a well-resourced and well-advised industry to attempt to make use of this legal provision. If political decisions are made to look like scientific decisions, accountability is lost, as the real reasons for decisions are never provided, and can therefore never be challenged. Science-based knowledge is clearly not the sole legitimate basis for decisions, but there is a danger that 'safe science' will be used more as a refuge than as a tool of understanding, as a way to escape the difficult political judgments and arguments.

Free Movement of Goods

Once a GMO has been authorised in the EU system, it enjoys the protection of internal market law, and free movement of goods throughout the EU.[135] Once authorised, and subject to conditions in that authorisation, a seed can in principle be grown anywhere in the EU, and food can in principle be sold anywhere in the EU. The possibility for autonomous Member State action after Community action is found in the 'safeguard clauses' contained in the legislation and in Article 95 of the Treaty itself. In both cases, however, the scope of national independence is restricted.

To begin with the safeguard clauses, Directive 1990/220 provided that a Member State with 'justifiable reasons' to consider that a product 'constitutes a risk to human health or the environment' could 'provisionally' restrict or prohibit its national use and/or sale.[136] Quite what would have constituted a justifiable reason was not clear, although the Commission was of the view that new scientific evidence would be required. In support of its safeguard measures on the maize at issue in *Greenpeace*, Austria had provided (along with a scientific report rejected as not constituting new scientific evidence) an opinion poll signed by one fifth of the population.[137] The Commission took the view that Austria had no 'justifiable reason' for its measure.

The new Deliberate Release Directive clarifies the need for a scientific basis for safeguard measures, allowing a Member State provisionally to restrict use and/or sale of a GMO only where there is 'new or additional information . . . affecting the environmental risk assessment or reassessment of existing information on the basis of new or additional scientific knowledge' and it has 'detailed grounds for considering that a GMO . . . constitutes a risk to human health or the environment'.[138] The narrow basis for Member State safeguard action not only emphasises the centralisation of authority but also

---

135 This is explicit in Directive 2001/18, above n. 10, Article 22.

136 Directive 1990/220, above n. 5, Article 16, see *Greenpeace*, above n. 24.

137 *Greenpeace,* above n. 24. See the discussion in Hervey, above n. 47, p. 324.

138 Directive 2001/18, above n. 10, Article 23. Article 20 provides for Community action if new information becomes available that applies to the whole EU.

reinforces the conclusion that science provides the primary source of legitimacy on GMOs. A Member State decision to take safeguard measures is notified to the Commission and other Member States, and a Commission decision is taken on the matter, with Council involvement through comitology.[139] As has been the case with all post-moratorium authorisations so far, if the Member States fail to agree, the Commission acts. The Commission was nevertheless reluctant to take action against national measures at the height of the controversy over GMOs. And apparently rightly so, because in 2005 the Council actually defeated, by qualified majority, a Commission proposal to request the withdrawal of Austrian safeguard measures in respect of certain GM maize. In response, the Commission consulted EFSA on these measures for a second time. Following EFSA's conclusion that it could still see no reason to believe that the continued placing on the market of this GMO would have adverse effects, the Commission resubmitted its proposals to Council. Again, a qualified majority rejected the Commission's proposals, challenging the failure to reassess the GMO in question under the 2001 Deliberate Release Directive, and arguing also that 'the different agricultural structures and regional ecological characteristics in the European Union need to be taken into account in a more systematic manner in the environmental risk assessment of GMOs'.[140] This suggests at the very least that the Member States are still not content with the regulatory structure applied to what remains an acutely contentious and sensitive issue. Responding to the concerns expressed in Council, the Commission's next proposal on the Austrian safeguard measures targeted only the food and feed aspects (and so not the cultivation) of the GMO. Council was this time unable to reach a qualified majority in either direction, leaving the Commission free to adopt its proposal.[141]

---

[139] Note that the European Parliament Committee on Environment, Public Health and Food Safety would have applied the regulatory procedure with scrutiny, above n. 8, to safeguard clauses; see Report 23 July 2007. This was not picked up in the text adopted by the European Parliament, 14 November 2007.

[140] 2773rd Council meeting, 18 December 2006, rejecting European Commission, *Proposal for a Council Decision concerning the provisional prohibition of the use and sale in Austria of genetically modified maize (Zea mays L. line T25) pursuant to Directive 2001/18/EC* COM (2006) 510 final. See also 2785th Council meeting, 20 February 2007, rejecting European Commission, *Proposal for a Council Decision concerning the provisional prohibition of the use and sale in Hungary of genetically modified maize (Zea mays L. line MON810) expressing the Bt cryIA(b) gene, pursuant to Directive 2001/18/EC* COM (2006) 713.

[141] European Commission, *Proposal for a Council Decision concerning the provisional prohibition of the use and sale in Austria of genetically modified maize (Zea mays L. line T25) pursuant to Directive 2001/18/EC* COM (2007) 589; European Commission, *Proposal for a Council Decision concerning the provisional prohibition of the use and sale in Austria of genetically modified maize (Zea mays L. line MON810)*

Member States have also turned to the safeguard provisions that apply to the 'common seed catalogue'. To be cultivated in the EU, a GM seed variety, like any other seed variety, needs to be registered in the EU's common seed catalogue, and hence needs to be 'distinct, stable and sufficiently uniform and . . . of satisfactory value for cultivation and use'.[142] This legislation provides somewhat more generous grounds for safeguard action. A Member State may seek authorisation (from the Commission, subject to comitology) to prohibit or make conditional the use of a variety when 'it is well known that the variety is not suitable for cultivation in any part of its territory because of its type of maturity class'.[143] Poland has successfully taken advantage of this provision when 'climatic and agricultural factors' were found to pose a 'permanent obstacle' to cultivating particular varieties in Poland.[144] The Directive also contains more familiar criteria of 'plant health', or a 'risk for the environment or for human health', and, although there is no explicit reference to a need for scientific evidence, we might expect these problems to be scientifically established.[145] Greece attempted to prohibit a number of GM seeds from its territory under this provision, claiming that it was for the protection of the 'rural environment'. However, when the Commission sought clarification, Greece said that the adverse effects it was concerned about were of an economic nature. The Commission, not surprisingly, denied authorisation on the basis that 'none of the specific provisions' of the Directive were satisfied.[146]

The relationship between the safeguard clauses in the Deliberate Release Directive and the Food and Feed Regulation is a little ambiguous. The Deliberate Release Directive cedes to sectoral legislation only if it contains 'a safeguard clause at least equivalent to that laid down in this Directive',[147] and

---

pursuant to Directive 2001/18 COM (2007) 586, and 2826th Council meeting, 14 November 2007.

142 Directive 2002/53/EC on the common catalogue of varieties of agricultural plant species [2002] OJ L 153/2, Article 4; Directive 2002/55/EC on the marketing of vegetable seed [2002] OJ L 193/33, Article 4 (although without the 'satisfactory value for cultivation and use' requirement, other than in respect of industrial chicory).

143 Directive 2002/53, above n. 142, Article 16(2).

144 Commission Decision 2006/335 authorising the Republic of Poland to prohibit on its territory the use of 16 genetically modified varieties of maize with the genetic modification MON810 listed in the Common catalogue of varieties of agricultural plant species, pursuant to Council Directive 2002/53/EC [2006] OJ L 124/26.

145 Directive 2002/53, above n. 142, Article 18(2).

146 Commission Decision 2006/10 concerning the provisional prohibition in Greece of the marketing of seeds of maize hybrids with the genetic modification MON810 inscribed in the common catalogue of varieties of agricultural plant species, pursuant to Directive 2002/53/EC [2006] OJ L 7/27, after Council failed to reach a qualified majority in either direction.

147 Directive 2001/18, above n. 10, Article 12(1).

the Food and Feed Regulation provides that the Deliberate Release Directive's safeguard clause shall not apply to authorisations granted under that Regulation.[148] But it is debatable whether the new safeguard clause is 'equivalent', as it even more seriously restricts the possibility of autonomous safeguard action. 'Emergency measures' may be triggered either by an opinion from EFSA, or where 'it is evident that products . . . are likely to constitute a serious risk to human health, animal health or the environment'.[149] The standard seems high: measures are concerned only with 'serious risk', and 'evident' seems to assume a high level of proof. Once emergency measures are triggered, the centralised procedure set out in the Food Safety Regulation applies, providing for Commission action through comitology.[150] Autonomous Member State action is possible only where the Member State has informed the Commission of the need to take emergency measures and the Commission has not acted in accordance with the Regulation. Again, the measures go to comitology.

Article 95(4) and (5) of the EC Treaty provide an alternative route for autonomous Member State action. These provisions allow derogations from measures adopted (such as the GMO legislation[151]) on the basis of Article 95(1), which provides for the harmonisation of laws for the 'establishment and functioning of the internal market'. Paragraphs (4) and (5) were designed to reassure certain Member States that the move away from unanimity in Council for internal market measures would not compromise their high standards of protection in areas such as the environment and public health. Whilst safeguard clauses are really practical only in respect of individual GMOs, on a case-by-case basis, Article 95(4) and (5) might apply more generally. Measures must be notified to the Commission, which can accept or reject them. The two paragraphs take slightly different approaches. Article 95(4) allows for the maintenance of existing national measures on the 'grounds of major needs referred to in Article 30,[152] or relating to the protection of the environment or the working environment'. Article 95(5) allows for the introduction of new national measures for the protection (only) of 'the environment or the working environment'. A 'specific problem on public health in a field

---

148 Regulation 1829/2003, above n. 12, Article 5(5).

149 Regulation 1829/2003, above n. 12, Article 34. See also Article 10 on new information affecting the authorisation.

150 Regulation 178/2002, above n. 25, Articles 53 and 54.

151 Regulation 1829/2003, above n. 12, is also based on Articles 37 and 152(4)(b).

152 So on the grounds of public morality, public policy or public security; the protection of health and life of humans, animals or plants; the protection of national treasures possessing artistic, historic or archaeological value; or the protection of industrial and commercial property.

which has been the subject of prior harmonisation measures' can be brought to the attention of the Commission under Article 95(8), and the Commission must 'immediately examine' whether to propose an 'adaptation' of those measures.

Article 95(5) is more strict than Article 95(4). New measures are thought more likely to jeopardise harmonisation, because existing measures will have been taken into account in the harmonisation.[153] Article 95(5) requires that new measures be 'based on new scientific evidence relating to the protection of the environment or the working environment on grounds of a problem specific to that Member State, arising after the adoption of the harmonisation measure.' Austria notified under Article 95(5) measures banning the use of GMOs in the Upper Austria region. The measures were concerned with the protection of organic agriculture, as well as protection of nature and the environment. They were rejected by the Commission on a number of grounds, including failure to provide new scientific information.[154] The question of the 'novelty' of scientific information is central to the operation of Article 95(5).[155] Austria relied on a report published after adoption of the legislation, but the data in the report 'were for a large part available prior to the adoption of Directive 2001/18/EC' and 'the vast majority of the sources were published prior to the adoption of Directive 2001/18/EC'.[156] Demanding brand new *data*

---

153 See for example Case C-512/99 *Germany v Commission* [2003] ECR I-845.

154 Commission Decision 2003/653 Relating to National Provisions on Banning the Use of Genetically Modified Organisms in the Region of Upper Austria notified by the Republic of Austria pursuant to Article 95(5) of the EC Treaty [2003] OJ L 230/34. In a more recent Article 95(5) decision, the Commission rejected a Polish application on the basis of a failure to provide new (or any) scientific information. This decision, especially the breadth of Polish concerns and the speed with which they are dismissed, illustrates clearly the narrow scope of the Article 95(5) derogation. See Commission Decision 2008/62 relating to Articles 111 and 172 of the Polish Draft Act on genetically modified organisms, notified by the Republic of Poland pursuant to Article 95(5) of the EC Treaty as derogations from the provision of Directive 2001/18 [2008] OJ L 16/17.

155 Hanna Sevenster, 'The Environmental Guarantee After Amsterdam: Does the Emperor Have New Clothes?' (2000) 1 *Yearbook of European Environmental Law* 281, argues for a purposive (environmental protection) approach; Nicholas de Sadeleer, 'Procedures for Derogations from the Principle of Approximation of Laws under Article 95 EC' (2003) 40 *Common Market Law Review* 889. De Sadeleer argues that the requirement for 'new' scientific evidence cannot be taken too literally, p. 901.

156 Decision 2003/653, above n. 154, para. 65. The Commission relies heavily on opinions of EFSA. Another key element of the Commission's decision was that the protection of organic farming was deemed to be a socioeconomic, rather than an environmental, issue; see Chapter 4 below. Chapter 4 also discusses the role of Directive 1998/34/EC Laying Down a Procedure for the Provision of Information in the Field of Technical Standards and Regulations [1998] OJ L 204/37.

places a very heavy burden on the Member State. Whilst it was not concerned with Article 95, the CFI decision in *Artegodan*, on the EU's withdrawal of authorisation of certain diet pills, was similarly strict. There was no new scientific evidence on diet pills, only an alleged new consensus on the lack of benefits of the pills. This was held by the CFI to be an insufficient basis for the withdrawal of the pills, which is justified:

> only where a new potential risk or the lack of efficacy is substantiated by new, objective, scientific and/or medical data or information . . . the application of a new assessment criterion, which reflects a current consensus in the medical community, is justifiable during the period of the authorisation's validity only if that development is based on new data or information.[157]

The stringency of this approach to 'new scientific information' is obvious when one looks at Article 95(4), which applies when a Member State wishes to maintain national provisions after harmonisation, and which does not require 'new' scientific information. The Court held in *Denmark* that a Member State can rely on its different assessment of the risk to public health:

> In the light of the uncertainty inherent in assessing the public health risks posed by, *inter alia*, the use of food additives, divergent assessments of those risks can legitimately be made, without necessarily being based on new or different scientific evidence.[158]

Importantly, Advocate General Sharpston took a similar approach when the Austrian GMO case was appealed to the ECJ. She concluded that 'new conclusions drawn from existing data *may* constitute new scientific evidence within the meaning of Article 95(5) EC'.[159] Austria could not provide such evidence in this case, but this more generous approach is a realistic understanding of what might provoke a change in national policy, and would be a powerful start in efforts to allow for national diversity. The safeguard clause in the Deliberate Release Directive, which as noted above allows for the 'reassessment of existing information on the basis of new or additional scientific knowledge' moves towards this, but falls short in the demand for new or additional scientific knowledge.

---

157 *Artegodan*, above n. 69, para. 194. Unless there is a 'detailed acknowledgement' that an incorrect assessment had been made in the first place, ibid.

158 *Denmark*, above n. 114, para. 63. The Danish Government notified its provisions on sulphites, nitrites and nitrates in foodstuffs.

159 Case C-439/05 *Land Oberosterreich and Austria v Commission*, not yet reported in the ECR, para. 124. This point was not pursued by the ECJ. Note that in *Monsanto*, above n. 56, the Advocate General took the position that the moratorium falls within the scope of new information or a reassessment of existing information (explicitly allowed in the safeguard clause at issue in that case), para. 151.

The stringency of the 'specific to the Member State' criterion in Article 95(5) is also a little unclear. Citing EFSA, the Commission held that Austria had failed to establish that 'small structured farming systems' are specific to this region, or that 'this area of Austria has unusual or unique ecosystems that required separate risk assessments from those conducted for Austria as a whole or for other similar areas of Europe'.[160] The ECJ provided an important clarification when it reminded us that the requirement is not that the problem be 'unique' to the Member State, but that it be 'specific'.[161] The question remains as to how extensive the problem might be: in the words of Advocate General Sharpston, 'a specific problem clearly lies somewhere between one which is unique and one which is common, generalised or widespread'.[162] Both the CFI and the ECJ rather abruptly rejected Austria's appeal on the basis of lack of evidence identifying a problem specific to the Member State.[163] Neither the Commission nor the Courts addressed the argument that the Austrian level of commitment to organic farming is what makes the problem 'specific' to Austria. This may be because, as discussed in Chapter 4, the protection of organic farming is a concern (controversially) deemed to relate 'more to a socio-economic problem than to the protection of the environment or the working environment'.[164] Article 95(5) allows for the pursuit of only a very narrow range of objectives.

In its decision on the Austrian biotechnology measures, the Commission, with the ultimate approval of the CFI and the ECJ, relied very heavily on EFSA opinions, and focused its decision very closely on EFSA's view of the science. The Commission took rather a different approach in its defence of

160 Commission Decision 2003/653, above n. 154, para. 71.

161 Case C-439/05, above n. 159, paras 65–8. Sharpston AG had found that, in referring to a unique rather than a specific problem, the CFI had used the wrong legal criterion to decide the case. She examined in detail the different language versions of the different documents – the EFSA opinion was drafted in English, the official version of the Commission decision was in German, and the CFI decision was drafted in French. The ECJ by contrast found that the CFI had not misconstrued the meaning of 'specific' since it did not hold that a unique problem was required.

162 Case C-439/05, above n. 1598, para. 110.

163 T366/03 and 235/04 *Land Oberosterreich and Austria v Commission* [2005] ECR II-4005, para. 68; Case C-439/05, above n. 159. Neither Court considered all conditions of Article 95(5), which are cumulative.

164 Commission Decision 2003/653, above n. 154, para. 67. Note also that Cyprus's attempt to notify a measure requiring supermarkets to keep GM foods on shelves separate from non-GM foods on consumer protection grounds was simply not admissible under Article 95(5), Commission Decision 2006/255 concerning national provisions imposing on supermarkets an obligation to place genetically modified foods on separate shelves from non-genetically modified foods, notified by Cyprus pursuant to Article 95(5) of the EC Treaty [2003] OJ L 92/12.

Austrian safeguard measures before the WTO panel on the subject, discussed in Chapter 6, where it emphasised the importance of a contextual and sensitive approach to national regulatory objectives. This is not just an amusing bureaucratic about-face, although it is understandable that the Commission will take a different approach in different contexts.[165] It goes to the heart of the level of independence that should be granted in a world/EU trade context. The emphasis throughout the EU procedures, safeguard and Treaty provisions, is very much on scientific evidence.

Both Article 95 and the safeguard provisions are subject to the precautionary principle. Indeed, the ECJ views safeguard clauses 'as giving specific expression to the precautionary principle'.[166] The same familiar if inchoate criteria for the use of the precautionary principle are applied as at EU level: protective measures 'may not properly be based on a purely hypothetical approach to risk, founded on mere suppositions which are not yet scientifically verified', and measures have to be 'based on a risk assessment which is as complete as possible'.[167] The emphasis remains on the scientific evidence, in order to police the legitimacy and objectivity of regulatory decisions. We might also note that there is a greater capacity and willingness in the Commission's administrative procedures to subject the science relied on by the Member States to 'centralised' scrutiny than there is in judicial review of EU action. In the Austrian decision, EFSA assessed the Austrian science from scratch, and the Commission (with two layers of Court approval) relied on EFSA's dismissal of the competing science. By contrast, the Courts control EU discretion by reference to a particular type of knowledge (science), but engage far less with the content of the science.[168]

The ability of Member States to block the commercialisation or cultivation of authorised GM agriculture or food in their territory is focused on the science of risk to health or the environment. Agricultural biotechnology, as

---

165 Sara Poli, 'The EU Risk Management of Genetically Modified Organisms and the Commission's Defence Strategy in the Biotech Dispute: Are they Inconsistent?' in Francesco Francioni and Tullio Scovazzi (eds), *Biotechnology and International Law* (Hart Publishing, 2006), examines in detail the inconsistencies between the Commission's internal and external stances on GMOs.

166 *Monsanto*, above n. 56, para. 110, citing *Greenpeace*, above n. 24.

167 *Monsanto*, above n. 56, paras 106 and 107.

168 Veerle Heyvaert, 'Facing the Consequences of the Precautionary Principle in European Community Law' (2006) 31 *European Law Review* 185, pp. 196–7, argues that the CFI imposes 'a burden of production rather than a far more onerous burden of persuasion', and that the Court's assessment of the quality of the science is 'perfunctory'. Almost by default then, national measures (unless they reach the Court by preliminary reference rather than administrative process) are subjected to greater scrutiny than EU measures.

discussed in Chapter 2, raises a host of other difficult social questions. These are very likely to manifest themselves differently around the Member States, and there may even be a strong cultural element to the politics of biotechnology.[169] The sensitivity of national preferences is clearly recognised in Council's political support for national safeguard measures. But, on the face of the law, the scope for engagement with citizen values or preferences in the Member States after authorisation is virtually non-existent. Some greater autonomy should be possible. Allowing for Member States to take divergent views of scientific evidence, as discussed above,[170] would be a start, with regulators reaching different conclusions in different social contexts, for example demanding more certainty on safety in respect of a particularly sensitive risk.

More ambitiously, applying something like the 'other legitimate factors' criterion that applies at EU level to national derogations would at least begin to recognise the delicacy of the situation. There is an irony perhaps in advocating the extension of a principle that is likely to be difficult to apply even at EU level, and putting the law in place would obviously be only the beginning of a process. But 'other legitimate factors' is more than a rhetorical flourish, and as at EU level it would be a small step towards recognising the complexity of decision making on GMOs. It would probably be rarely used, but would allow space for distinctive responses around the EU. And this proposal would not be wholly out of step with the legal framework. Reference by the courts to consumer confidence and public concern suggest potential for openness, as does the possibility of considering the *need* for a genetic modification, at least as a factor in risk assessment. Moreover, the language of the legislation ('Member States may take into consideration ethical aspects when GMOs are deliberately released or placed on the market as or in products'[171]) and the policy ('the ethical acceptability of some areas of biotechnology is related to the diversity among Member States and is governed by national law in accordance with the principle of subsidiarity'[172]) are directly supportive of such an expansion.

Expanding the grounds of national autonomy to include broader political/democratic concerns is, of course, a hugely difficult proposal. Any simple response to 'public opinion' is vulnerable to abuse, giving Member States a free hand to avoid their obligations when they feel like it. This runs contrary

169 See Jasanoff, above n. 3. Although Jasanoff's attribution of cultural attributes along national lines is controversial, she convincingly demonstrates persistent and profound divergences in respect of biotechnology.

170 See also Chalmers, above n. 37.

171 Directive 2001/18, above n. 10, Recital 9.

172 *Council Conclusions and Roadmap*, above n. 1, p. 12.

to decades of development of internal market disciplines, and risks undermining the EU's primary achievement. But a democratic government should at least be able to contemplate representing democratically expressed concerns, whether those concerns are about safety, socioeconomics or ethics. Without suggesting that such an extensive process should always be required, something like *GM Nation?*, discussed in Chapter 2, allows a national government to explain and defend, in a way that permits challenge, the reasons for particular decisions beyond a scientific assessment of protection of the environment and health. There are problems with the EU involving itself in these rather profound issues of national democratic processes, but, given that it *is* involved, it might turn out to be less objectionable for it to look in detail at how the Member State has ascertained what constitutes an 'other legitimate factor' than to demand a sound science basis and narrow range of objectives for all decisions. And whilst expanding the range of values that can justify interruptions to free trade is challenging, so are the alternatives. At the moment, the law continues to malfunction. Continued legal disobedience causes problems at least as great as a more flexible approach. And it would be just as problematic in terms of the reputation and legitimacy of the internal market and the EU for the centre to 'win' this legal battle and for GMOs to be pushed onto reluctant national markets. Turning to location-specific authorisation of GMOs, just as polluting installations are authorised at the local rather than the central level, provides no easy answers, because their status as products makes GMOs especially sensitive to internal market disciplines. Expanding the basis of national autonomy is a less radical change, and fits well with the framework of the policy and case law.[173]

Clearer criteria for a more generous approach to derogations from common action would enable greater autonomy for Member States that wish to reflect the democratic will. Given that there is scope and an apparent willingness for Member States to show a certain amount of sympathy when they are asked to assess each other's safeguard clauses, it would also subject the de facto, somewhat chaotic, autonomy that is already being exercised to a plainer discipline.

Multi-level Governance and the Politics of GM

Authority on GMOs is disputed in at least two main directions, that is, national/central authority and political/scientific authority. The legislation attempts to mediate between these different pressures. The contest between scientific and political authority appears formally to have been won by the

[173] Chalmers argues for greater national autonomy in respect of food generally, above n. 4.

politics – politicians have the final say, and can use a range of political considerations (other legitimate factors) to justify their decision. But the authority of science as a tool of decision making in practice and in law remains very, even overwhelmingly, powerful, and the Commission's decisions suggest that it prefers to follow EFSA opinions. The relationship between national and EU authority is equally complex. But, at least in legal form, agricultural biotechnology has been more or less successfully 'Europeanised' as a policy area.[174] This Europeanisation of GMOs remains controversial, and the legislative and administrative arrangements contain a complex mix of concessions to national priorities and autonomy, as discussed throughout this chapter. It is in any event impossible to maintain a strict dichotomy, if only because Member States and national interests constitute some of the key EU institutions and are represented in others. And whilst there is no wholehearted embrace of the tools of 'new' governance in the regulation of GMOs,[175] but a reliance largely on longstanding mechanisms of decision making (comitology, scientific advice), there are considerable efforts within and beyond the legislation to expand the space available for informal debate before positions become entrenched. The networking of EFSA, the circulation of applications and views, and the building-in of time for debate and discussion all call on these alternative approaches.

The most significant impetus to the apparent settling of authority at EU level must be the demands of the internal market, as discussed in the previous section. As the capacity and ambitions of the EU expand, we perhaps need to remind ourselves how central the internal market is to the purposes of the EU. Were it not for concern about barriers to trade, there would be little reason why Member State disagreement over GMOs could not be left to play out on the ground.[176] But the free movement of goods (including those produced by biotechnology) is not just important in its own right, or even simply as an instrument of economic prosperity. Since the birth of the European Economic Community, economic integration has been both a symbol and a precursor of increasing political integration. Different national approaches to the regulatory challenge of GMOs raise barriers to the completion of the internal market, and in turn threaten the core of the EU's ambitions. The international context also contributes to the importance of the internal market in this policy area.

---

174 Jasanoff, above n. 3, sees biotechnology policy as an important element of Europeanisation, and see generally the *Life Sciences Strategy*, above n. 1.

175 See especially Gráinne de Búrca and Joanne Scott (eds), *Law and New Governance in the EU and the US* (Hart Publishing, 2006).

176 See also the discussion in Chapter 4 of the 'Europeanisation' of coexistence, notwithstanding the potential for considerable national flexibility under Article 26a of Directive 2001/18, above n. 10.

External pressures from trade partners, particularly through the WTO, amplify the importance of maintaining internal market disciplines. And the interest of the EU, a powerful economy, in a strong and predictable international trade regime magnifies the Commission's determination in this matter.

EFSA's role in the authorisation process is a visible centralisation of authority. However, even this is not a simple allocation of power. Networking is an intrinsic part of EFSA's legal constitution, bringing the Member States into the process and implying at least the openness of scientific views to critique from national technical experts. The taking of decisions (albeit *in extremis*) by the Commission in the years following the end of the moratorium is also a centralisation of authority. But again, whilst the final political decision is taken by an EU institution, all Member States are involved through comitology. EU lawyers are familiar with the arguments of Joerges and Neyer that comitology is a forum for 'deliberative supranationalism',[177] where deliberation, persuasion and practical reason mean that individual and national interests are escaped, to respond to the unique demands of EU decision making. By comparison with the failure of nation states alone to take external interests into account, this provides EU decision making with a new democratic dignity.[178] However, the type of involvement in either the comitology or the EFSA 'networking' committees is far from transparent to the relevant publics, and elite views are better represented than popular views. Without wishing to oversimplify, the political questions tend to be more apparent at the national level, which is more susceptible to democratic pressures. But the reflection of public opinion through various committees or in Council is problematic. Even with the best will of national governments, it is striking how difficult it would be to represent the views of the public as expressed in a public exercise such as the UK's *GM Nation?* debate through comitology. Not only is the UK only one Member State of 27, but the *results* of the consultation were complex and nuanced, and only with great difficulty if at all fitted into the framework of the legislation. Presumably each Member State faces similar difficulties. If public participation in this area is at least in part about the representation of public opinion to decision makers, we could attempt to fill this gap at EU level. Representation of public concern at EU level is difficult, however, if only because European public opinion at this level is associated more or less

---

177 Christian Joerges and Jurgen Neyer, 'From Intergovernmental Bargaining to Deliberative Process: The Constitutionalisation of Comitology' (1997) 3 *European Law Journal* 273.

178 Christian Joerges, 'Deliberative Political Processes Revisited: What Have We Learnt About the Legitimacy of Supranational Decision-Making?' (2006) 44 *Journal of Common Market Studies* 779. This idea of external accountability is discussed in more detail in the WTO context in Chapter 6.

entirely with pan-European or international interest groups.[179] These groups are hugely important for the critical examination of expert claims, but nevertheless potentially distant from local concerns. In these circumstances, greater national autonomy is highly appropriate.

Whatever the difficulties, it must be accepted that the legislation makes efforts to compromise between national and central authority, just as it makes efforts to move beyond a purely science/risk basis for decisions.[180] Moreover, the EU is a powerful way of requiring states to take account of external interests. If there is ultimately no happy consensus, the provision of a voice to Member States, at both the 'science' and the 'political' stages, should at least contribute to the national acceptability of decisions with which a particular Member State and its citizens disagree. This does not, however, seem always to be the case. Member State involvement in authorisation decisions is weak whenever division means that the Member States cannot act collectively.

In these circumstances, greater deference to national difference on this issue would be beneficial. A recognition of the profound divisions between the Member States, together with their democratic responsibilities in a matter of national importance, could justify divergent national conclusions on the release of GMOs. The reasons for Europeanisation of this policy area are far from trivial, and will be insisted upon, in particular the principles of multilateralism and free movement that characterise the EU. But the grim insistence on EU authority has the potential to undermine the perceived legitimacy of the EU institutions, making the current path just as risky as enhancing national autonomy. The EU institutions cannot assume that European publics will simply become accustomed to the commercialisation of GMOs. Somewhat more generous grounds for safeguard or Article 95 measures, as discussed above, may be worth exploring. Because the central authority on this subject remains fragile, there is a danger that the EU will show itself to be incapable of imposing (or even identifying) its will.

CONCLUSIONS

The EU institutions have (for now) successfully defined policy towards GMOs as an 'EU' issue, so that conditions for marketing and use are determined at that level. There is a whole range of pressures pointing towards the encouragement

179 See Lee, above n. 12, Chapter 5.

180 Sebastian Krapohl, 'Credible Commitment in Non-Independent Regulatory Agencies: A Comparative Analysis of the European Agencies for Pharmaceuticals and Foodstuffs' (2004) 10 *European Law Journal* 518, argues that Member States retain strong control through comitology.

and facilitation of development of agricultural biotechnology in the EU, which compete with pressures exerted by those opposing the technology, either absolutely or in particular circumstances. The organisation of the legislation attempts to tread a delicate path between national and central authorities and between reaping the benefits and protecting public interests, as well as between scientific and political understandings of GMOs. The sensitivity of the balance is emphasised by the uncertainty that pervades our understanding of effects in this area, not to mention public awareness of that uncertainty.

The Commission is apparently content that the moratorium is over. The ending of a sensitive and high-profile moratorium by the Commission in the face of such profound division is, however, politically provocative. And it by no means proves the success of the new legislative framework. Indeed, there are still problems with basic implementation in the Member States, let alone with pushing products through the authorisation process.[181] Moreover, EU producers and retailers continue to reject GM ingredients,[182] and Member States continue to apply dubious (in legal terms) safeguard measures. But the Commission seems to see the ultimate embrace of GMOs as inevitable:

> The technology and its applications are developing rapidly – the Commission believes that Europe's policy choice is, therefore, not whether but how to deal with the challenges posed by the new knowledge and its applications.[183]

And the perception that the EU is somehow missing out on a wonderful opportunity explains much of the ambivalence towards messy politics in this area:

> Uncertainty about societal acceptance has contributed to detracting attention in Europe from the factors that determine our capacity for innovation and technology development and uptake. This has stifled our competitive position, weakened our research capability and could limit our policy options in the longer term.[184]

---

[181] It is an indication of the Commission's seriousness that on 12 December 2006 the Commission commenced further legal proceedings against France under Article 228 of the Treaty, seeking the imposition of a fine for the failure to comply with an earlier ECJ judgment that France had inadequately implemented the 1990 Directive, Case C-121/07 *Commission v France*. Not yet reported in ECR.

[182] See for example COM (2006) 626 final, above n. 32, which states that 'few food products labelled as genetically modified are at the present time on the Community market', because of 'factors that are not related to the legislative framework, such as consumer demand and the policies of food producers and retailers', para. 3.1.

[183] *Life Sciences Strategy*, above n. 1, pp. 8 and 9.

[184] *Life Sciences Strategy*, above n. 1, p. 8.

Treading a course between public protection and economic progress is, of course, what all regulators try to do. The challenges for the EU are particularly pronounced. Faced with endemic scientific uncertainty, mistrustful publics, problematic legitimacy, and an awkward international trade context, the EU attempts to steer a course between deeply divided views. The legislation provides a rigorous mechanism by which to assess, contest and debate the safety of GMOs. But as a social 'problem', agricultural biotechnology raises a whole range of concerns, including but certainly not limited to scientific questions of safety for the environment and human health. This breadth of concerns poses a challenge to a system of regulation much more comfortable with relying on scientific assessments and the appearance of objectivity and independence that they provide. The regulation in place is certainly 'more than a technocratic regime',[185] including a clearly political phase of decision making. The real depth of the political enquiry is still to be seen, and the ambiguity is enhanced by the extreme confidence that the Commission demonstrates in the scientific opinions actually before it.

Simply saying that all decisions are political decisions is perhaps trite, but it is true that very few decisions can be purely technical, amenable to solutions by the neutral, value- and judgment-free application of expertise. So the basis for their resolution benefits from normative debate. In part because the political concerns are far more visible in the national democratic sphere than at EU level, there is considerable tension between national and EU authority on GMOs. And again, there is space in the legislation for national influence over decisions on GMOs. But again, that influence is ambiguous and limited. The mismatch between national politics and European decisions has led to open conflict, and is capable of doing so again.

Agricultural biotechnology raises profound questions about the capacity of states to respond to environmental/public health challenges and also to the desires of their citizens. It is important not to be pessimistic. The new set of legislation was a self-conscious attempt at compromise and confidence building, with great awareness of the fact that legitimacy cannot be taken for granted. The institutional and political resistance to stepping back from the regulation of GMOs has nevertheless led to a range of 'democratising' and 'decentralising' measures, including networking, openness and transparency, participation, the choice of comitology procedure. Perhaps most importantly, risk assessment is not the end of the analysis, and the political nature of the final decision is clearly recognised in principle, if difficult to put into place in practice. Nevertheless, accommodating the complexity of public views at the

---

185 Christine Landfried, 'The European Regulation of Biotechnology by Polycratic Governance' in Joerges and Vos, above n. 42.

EU level remains profoundly difficult, and I have argued here that there should be greater deference to national decisions. This poses enormous challenges, but it is important that the Member States be allowed at least to contemplate an autonomous response to strong views of their own citizens. To the extent that they can already do so, clearer criteria would, as well as enhancing the legitimacy of EU decision making, both enable democratic responsiveness on the part of the Member States and discipline recourse to trade-disruptive measures.

4. Living with GMOs (1): coexistence, liability and labelling

INTRODUCTION

The regulatory story of GMOs does not begin or end with authorisation.[1] GMOs are cultivated and marketed in a particular legal context, and that context provides the social conditions for the development of the technology. The suitability of the rules applying to GMOs after authorisation is likely to feed into the acceptability of authorisation, and indeed authorisation decisions are incomplete if they are not made in the context of a reasonably predictable post-authorisation framework. But at the moment, that framework is difficult to pin down. This chapter examines three important and interrelated aspects of the legal environment in which GMOs will be grown and sold in the EU: coexistence, liability and labelling. Chapter 5 continues this theme, examining the application of patent law to GMOs. It is common to see these questions presented as if they were purely technical legal issues, amenable to incremental development and neutral application by experts. In fact, they are subject to profoundly political choices, determining the distribution of costs, benefits, risks and uncertainties. As such they should be the subject of normative debate in exactly the same way as any other part of the regulatory framework.

Total isolation of GM material, certainly once agricultural biotechnology is widespread in the EU, is impossible. There will inevitably be some level of mixing between GM and non-GM material, through a variety of means, including natural cross-pollination by wind or insects, the survival of GM 'volunteers', and mixing by farm machinery, or in storage, distribution or processing. Those unhappy about 'gene drift' characterise the unwanted presence of GM material as 'contamination' or 'pollution'. 'Pollution' and 'contamination' are not neutral terms, and EU legislation and policy pointedly avoid this sort of language. The Commission rejects national use of the word

[1] And indeed authorisation contains some forward-looking elements including a monitoring plan, Directive 2001/18 on the Deliberate Release into the Environment of Genetically Modified Organisms and Repealing Council Directive 90/220/EEC [2001] OJ L 106/1, Article 20.

'contamination'[2] and generally prefers the term 'admixture'.[3] The Directive uses the somewhat laboured, not to mention deeply ambiguous, phrase 'adventitious or technically unavoidable' to qualify the unsought presence of GM material in non-GM products.[4] This language is as heavily value laden as 'contamination', trying hard to neutralise and depoliticise the issue. The language chosen in part reflects whether the presence of GM material out of place is deemed to be a problem. As with most other forms of pollution and contamination, environmental law attempts to make this question more susceptible to regulatory management by setting thresholds of acceptability for matter out of place, although the language of 'contamination' and 'admixture' is not used entirely consistently with these thresholds.[5] Further, whilst the *level* of such thresholds is always contentious, in the case of GM material, the *role* of the thresholds is equally problematic, as discussed further below.

The measures by which coexistence will be ensured are an important element of the overall system of regulation. The Commission's view is that the regulation of 'coexistence' between GM and conventional or organic farming is designed to keep levels of GM material in non-GM products to some acceptable level and to minimise any associated problems. According to the legislation, coexistence is a national competence. National policies are, however, seriously constrained by Commission policy and EU law more generally, without the centre taking on the responsibility of ensuring coexistence. As with authorisation, discussed in Chapter 3, the tight control being exercised by the centre in the interests of market freedoms could come at a high price. It rests on a narrow and contested understanding of what is at stake in the regulation

---

2 'Other terms would be more appropriate to describe the situation at hand'; see European Commission, Detailed Opinion under Directive 1998/34/EC Laying Down a Procedure for the Provision of Information in the Field of Technical Standards and Regulations [1998] OJ L 204/37, *Notification 2003/475/A (Genetic Engineering Precautionary Measures Act).*

3 See for example European Commission Recommendation, *Guidelines for the Development of National Strategies and Best Practices to Ensure the Co-Existence of Genetically Modified (GM) Crops with Conventional and Organic Farming* C (2003).

4 Regulation 1829/2003/EC on Genetically Modified Food and Feed [2003] OJ L 268/1, Article 12(2). See the discussion in Les Levidow and Susan Carr, 'GM Crops on Trial: Technological Development as a Real-World Experiment' (2007) 39 *Futures* 408.

5 In the Commission's Detailed Opinion, above n. 2, the disapproval of the language of contamination applies implicitly even if the level of GM material exceeds the threshold. The language of 'contamination' is used when those thresholds do not apply, for example in the presence of unauthorised GM rice in rice imports; see Commission Decision 2006/578 on emergency measures regarding the non-authorised genetically modified organism LL RICE 601 in rice products [2006] OJ L 230/8, updated by Commission Decision 2006/601 [2006] OJ L 244/27.

of GMOs, and it is argued in this chapter that the Commission's approach is in many respects misconceived.

The approach to liability for harms associated with agricultural biotechnology is intimately connected to the regulation of coexistence between different forms of agriculture, as farmers suffering a loss of market or difficult agricultural management problems are likely to seek recompense. The pervasiveness of GMOs creates the potential for conflict over the use of land, taking the politics of GM to the local level. Liability questions also go beyond impacts on neighbouring farmers. In particular, GMOs could have ecological impacts as GM material enters the natural environment, or GMOs may have unanticipated health impacts when consumed. Who is responsible if things go wrong should be a key element of the regulatory regime for any new technology. This constitutes a real gap in the regulatory framework for GMOs.

And, finally, the official discourse on GMOs is wrapped in a rhetoric of consumer choice, underpinned by an obligation to label GMOs. One of the unspoken values of the regulatory framework for GMOs is the assumption that the market is the best place for the exercise of choice, and indeed that ethical commitments are simply a matter of choice. GM food has been something of a *cause célèbre* for consumer power, with many major food retailers and processors rejecting GMOs in the 1990s. The *regulatory* expectation seems to be that any public concerns about GMOs that are not caught by the authorisation process can be adequately dealt with by facilitating choice in the market. It is extremely difficult to find a space for debate of non-safety issues in respect of GMOs, and the mandatory labelling of GMOs has the potential to make the market a locus, albeit imperfect, for the expression of collective, political values, as well as individual preferences. The regulation, however, rests on assumptions that need to be picked apart. As well as straightforward gaps in the labelling regime, the tendency of the EU to conceptualise the 'public' in GM matters as consumers incurs the danger of misrepresenting public responses to biotechnology, so that apprehension about biotechnology becomes a concern about the consumption of particular products.[6] This has real potential to impoverish analysis and trivialise choice: which brand of tomato ketchup, rather than fundamental questions about how we run our agricultural systems.

---

6 See Robin Grove-White, Phil Macnaghten and Brian Wynne, *Wising Up: The Public and New Technologies* (The Centre for the Study of Environmental Change, 2000), especially p. 32.

COEXISTENCE IN EU LAW

The starting point for coexistence is the view of the Commission that all forms of agriculture should be able to coexist: 'No form of agriculture, be it conventional, organic, or agriculture using GMOs, should be excluded in the European Union'.[7] Coexistence problems recur throughout the life of GM products, and GM material could be mixed with non-GM products in the field or during storage, transport or processing. This is not unique – many GM products are agricultural commodity products subject to enormous trade flows, and economies of scale are sought through bulk storage and transport. Coexistence measures are supposed to be a way to maintain choice for farmers, including presumably the choice to reject GMOs. The Commission outlines 'an open-ended catalogue' of possible coexistence measures in a non-binding Recommendation on the subject,[8] including the use of separation distances between crops, buffer zones, and coordination of crop rotation and pollination times, along with measures such as using separate machinery for GM and non-GM crops, and careful management of volunteers and other weeds. Such measures do not guarantee the absence of gene flow, but should be designed to minimise its extent. If these measures are combined with the thresholds (discussed below) beneath which a product containing GM material need not be labelled as such, the Commission's position is that many of the problems that might be associated with coexistence will be avoided.

Farm management measures designed to restrict gene flow may have costs. The 'general principle' in the Commission's Guidelines is that farmers introducing a 'new production type' should bear responsibility for implementing the farm management measures.[9] So, in the short term at least, GM farmers should bear the bulk of the costs, although the Commission insists that we remember 'farmers who introduce the cultivation of conventional or organic crops in an area where GM farming is already present'.[10] Moreover, the prac-

---

[7] European Commission Recommendation, above n. 3, Recital 1. More recently, see European Commission, *Report on the Implementation of National Measures on the Coexistence of Genetically Modified Crops with Conventional and Organic Farming* COM (2006) 104 final. The need for further guidance is to be reconsidered in 2008; see also European Commission, *Communication on the Mid-term Review of the Strategy on Life Sciences and Biotechnology* COM (2007) 175 final.

[8] Commission Recommendation, above n. 3, para. 3.

[9] Commission Recommendation, above n. 3, para. 2.1.7.

[10] European Commission, Detailed Opinion under Directive 1998/34/EC Laying Down a Procedure for the Provision of Information in the Field of Technical Standards and Regulations [1998] OJ L 204/37, *Notification 2005/610/A (Draft Upper Austrian Genetic Engineering Precautionary Measures Act)*, para. II(4); similar phrasing can be found in other detailed opinions.

tical maintenance of any formal preference for the allocation of costs to GM farmers is another matter. Imposing farm management costs on the GM farmer has the potential to make considerable demands of regulators, and the organic or conventionally farming neighbour has the greatest sense of urgency on these issues. In practice, the party with an interest in avoiding GM material is likely to take on the daily grind of ensuring coexistence. The Commission also emphasises that measures should be coordinated between farms to achieve 'lower costs than if all measures would have to be taken by one farm or a limited number of farms'.[11] That is in itself unobjectionable, but really here the concern should be with *distribution* as much as with the calculation of overall costs. This anticipates elements of civil liability, discussed below (pp. 127–41), in respect of the costs of, for example, changed crop management or the eradication of GM volunteers.

Multi-level Governance – Again

It seems to be broadly accepted that coexistence is not an issue that can simply be left to the market, but that it requires some form of organisation, if not government regulation. Coexistence is a Member State competence under Article 26a of the Deliberate Release Directive: 'Member States may take appropriate measures to avoid the unintended presence of GMOs in other products'.[12] The language is open, and in theory leaves room for distinctive national approaches to the balance between different forms of farming. As ever, however, the division of national and central authority on this issue is actually rather more complex than it appears. This is a theme that pervades this chapter, but it is worth outlining here some of the main incursions into what appears on the face of Article 26a to be complete national freedom of action. Member States are constrained both by the general legal context and by the specific Commission framing of coexistence.

First of all, Article 26a must be read alongside Article 22, which provides that Member States may not 'prohibit, restrict or impede the placing on the market of GMOs . . . which comply with the requirements of this Directive'.[13]

---

11 European Commission, Detailed Opinion under Directive 1998/34/EC Laying Down a Procedure for the Provision of Information in the Field of Technical Standards and Regulations [1998] OJ L 204/37, *Notification 2004/538/A (Viennese Genetic Engineering Precautionary Measures Act)*, para. II(2)(a); similar phrasing can be found in other detailed opinions.

12 Directive 2001/18, above n. 1, Article 26a, introduced by Regulation 1829/2003, above n. 4, Article 43(2).

13 Note that, in its Detailed Opinions under Directive 1998/34, the Commission interprets Article 19 as precluding further consent obligations by Member States.

This is arguably simply an expression of the principle of free movement of goods in EU law, and the apparent flexibility in Article 26a is in any event subject to the requirement that 'appropriate measures' be consistent with the regulatory framework and objectives and basic EU law, including internal market law. More specifically, and relevant to the way in which that general law might be applied, Member State autonomy is placed in a restrictive legal and policy framework by the Commission. As provided under the legislation,[14] the Commission has issued the Recommendation on Guidelines on coexistence mentioned above. Although recommendations are non-binding,[15] this provides a highly influential, albeit contestable, starting point for coexistence.

Some of the Commission's recommendations are mentioned above, including primarily farm management measures such as introducing buffer zones and cleaning machinery. One highly significant overarching aspect of the Recommendation, reiterated in much subsequent policy, and indeed law,[16] is the Commission's understanding of coexistence as a purely economic issue.[17] This understanding relies on the fact that any environmental or health issues (and presumably any 'other legitimate factors') were assessed in the authorisation process, with anything unexpected being dealt with through the safeguard process. This narrow approach to coexistence is not surprising, given the regulatory framework,[18] which means that, whilst coexistence is a national

European Commission, Staff Working Document, *Annex to the Report on the Implementation of National Measures on the Coexistence of Genetically Modified Crops with Conventional and Organic Farming* SEC (2006) 313, interprets Article 19 as meaning that 'restrictions on the cultivation of GM crops in specific areas on environmental grounds can apply only to those GMOs for which such restrictions have been laid down in the final consent'. Article 19 certainly provides for the possibility of conditions on authorisation, so the implication must be that the Commission is of the view that this area is exhaustively harmonised.

[14] Directive 2001/18, above n. 1, Article 26(a), second paragraph.

[15] National authorities and courts are bound to take them into account, especially when 'they cast light on the interpretation of national measures adopted in order to implement them or where they are designed to supplement binding Community provisions', Case C-322/88 *Grimaldi* [1989] ECR 4407.

[16] See the Opinion of Sharpston AG in Case C-439/05 *Land Oberosterreich and Austria v Commission*, not yet reported in the ECR, discussed further below and in Chapter 3.

[17] *Communication from Mr Fischler to the Commission: Co-existence of Genetically Modified, Conventional and Organic Crops* SEC (2003) 258/4; Commission Recommendation, above n. 3, Recitals 4 and 5; also COM (2006) 104, above n. 7.

[18] Sara Poli, 'Restrictions on the Cultivation of Genetically Modified Organisms: Issues of EC Law' in Han Somsen (ed.), *The Regulatory Challenge of Biotechnology: Human Genetics, Food and Patents* (Edward Elgar, 2007).

responsibility, risk regulation is subject to harmonised legislation. The definition of coexistence as an economic issue should not, however, be left unchallenged. It is more complicated than it appears, and directly implicates the freedom of action of the Member State.

Basic internal market law in the EU prevents the Member States using economic arguments to justify an interference with the free movement of goods.[19] Greater leeway could be available if coexistence measures were also about social regulation, including environmental protection[20] and protection of health[21] and consumers.[22] In the context of Austrian restrictions on the ownership of agricultural land, the European Court of Justice (ECJ) has even shown some sympathy to the 'social objectives' of

> preserving agricultural communities, maintaining a distribution of land ownership which allows the development of viable farms and sympathetic management of green spaces and the countryside as well as encouraging a reasonable use of the available land by resisting pressure on land, and preventing natural disasters.[23]

In this context of internal market disciplines, the application of the precautionary principle to economic issues is problematic.[24] It might be anticipated that, in an area of such scientific uncertainty, the precautionary principle could be employed in particular cases to justify (or even require[25]) strict regulatory action to protect non-GM farming. Whilst the absence of a clear role for the

---

19 Case C-120/95 *Decker* [1998] ECR I-1831, para. 39. In Case C-203/96 *Chemische Afvalstoffen Dusseldorp BV v Minister van Volkshuisvesting, Ruimtelijke Ordening en Milieubeheer* [1998] ECR I-4075, the ECJ rejected the argument that economic protection offered to a national industry by a restriction on exports was justified because that industry provided environmental benefits: 'aims of a purely economic nature cannot justify barriers to the fundamental principle of the free movement of goods', para. 44.

20 Case 302/86 *Commission v Denmark* [1988] ECR 4607.

21 Case C-41/02 *Commission v Netherlands* [2004] ECR I-11375.

22 Case C-12/00 *Commission v Spain* [2003] ECR I-459.

23 Case 452/01 *Margarethe Ospelt v Schlössle Weissenberg Familienstiftung* [2003] ECR I-9743, para. 39.

24 In Case C-132/03 *Ministero della Salute v Coordinamento delle Associazioni per la Difesa dell'Ambiente e dei Diritti degli Utenti e dei Consumatori (Codacons)* [2005] ECR I-4167, it was argued that the precautionary principle demanded that baby food should be labelled as containing GMOs even if the unintentional traces of GM material fell below the legislative threshold discussed below. This argument was rejected by the Court on the basis that labelling is about information rather than safety, and the precautionary principle has already been applied in authorisation.

25 See the discussion of the precautionary principle in Chapter 2, especially Cases T-74/00, T-76/00 and T-141/00 *Artegodan and Others v Commission* [2002] ECR II-4945.

precautionary principle cannot mean that absolute scientific certainty is required for coexistence measures, conceptualising that risk as economic may sideline a more evaluative perspective on uncertainty. A clear place for the precautionary principle would also pre-empt any self-serving efforts to demand ever more scientific evidence before action is taken.

In principle, even the measures advocated by the Commission in its Recommendation are capable of raising internal market issues. Member States trying to justify national or regional restrictions on GM agriculture for coexistence purposes will turn to a complicated array of legal provisions. It will be recalled that safeguard clauses and Article 95 of the EC Treaty, discussed in Chapter 3, focus on action to protect against specific risks to health and the environment. Coexistence measures are more likely to be assessed under Directive 1998/34 on technical regulations, which requires Member States to notify draft measures in the field of standards and technical regulations to the Commission.[26] Failure to notify measures renders them inapplicable and unenforceable against individuals.[27] Many of the measures advocated by the Commission in its Recommendation will need to be notified if implemented by the Member States. The draft is subject to a three-month standstill period during which it may not be adopted by the Member State.[28] During that period, the Commission and the Member States consider the draft, and may decide that the measure does not introduce barriers to the single market. Alternatively, comments may be sent when the draft raises issues of interpretation, or there is a need for details of the arrangements relating to implementation, or a Detailed Opinion will be issued if the draft measure appears to hinder the operation of the internal market, extending the standstill period for a further three months.[29]

A number of Member States have notified draft coexistence measures under Directive 1998/34, and the Commission's approach to these measures suggests that it intends to keep a very tight rein on the subject.[30] The Commission consistently reiterates certain key propositions: coexistence

---

[26] Directive 1998/34, above n. 2. Summaries of notifications can be found at http://ec.europa.eu/enterprise/tris/pisa/app/search/index.cfm (accessed December 2007).

[27] Case C-194/94 *CIA Security International* [1996] ECR I-2201.

[28] Directive 1998/34, above n. 2, Article 9.

[29] One further option is for the Commission to extend the standstill period to 12 months, if the proposed draft covers an area where the Commission proposes to legislate.

[30] Detailed Opinions are not publicly available, but I am grateful to the Commission for making Detailed Opinions available to me in respect of Notifications: 2003/475/A, 2004/459/A, 2006/73/P, 2005/634-637/HU, 2005/271/P, 2004/133/D, 2006/455/SK, 2004/538/A, 2005/005/A, 2003/200/A, 2005/610/A, and the following national responses: Communications: SG(2006) D 53080 on Notification 2006/

measures cannot address questions of environmental or human health protection; coexistence measures must be proportionate, which implies (inappropriately, as discussed below) use of the labelling threshold as a benchmark; administrative efforts and associated costs imposed on farmers cannot be such as to prevent cultivation of GM crops. The Commission's position is also clearly that Member States cannot require any additional authorisation process for local cultivation, although farmers can be required to provide information to a local regulator, as long as the information required is not 'disproportionate or . . . too difficult to comply with by an average farmer'.[31]

The Commission is vigilant to ensure that GM farming is not precluded by coexistence measures. But others advocate using coexistence measures to restrict GM agriculture to the greatest extent possible. Even those who are not in principle against agricultural biotechnology may seek high levels of protection for existing farming, and/or the maintenance of relatively uncontaminated farming. Accordingly we see a movement calling for 'GM-free' zones, or at least regimes highly restrictive of GM agriculture, and various regions have declared themselves to be GM-free.[32] GM-free zones might be imposed for a number of reasons. Their proponents frequently call on ideas of locally and culturally specific agricultural production. As opposition to the multinational corporations and the uniformity implied by agricultural biotechnology, this is a direct response to some of the concerns outlined in Chapter 2. GM-free zones could also have more direct environmental or public health objectives. The adequacy of risk regulation as a response to agricultural biotechnology is, as discussed in Chapters 2 and 3, disputed. Risk regulators explicitly may not take into account ignorance or 'hypothetical risk', and in these circumstances keeping some farming areas literally GM-free, or as close as is still possible,

---

455/SK; SG(2006) D/51855 on Notification 2005/634/HU, 2006/637/HU; SG(2004) D/51849 on Notification 2004/133/D; SG(2006) D/51039 on Notification 2005/610/A; SG(2005) D/52124 on Notification 2005/5/A; SG(2005) D/51017 on Notification 2004/538/A; SG(2005) D/51284 on Notification 2004/459/A; SG(2004) D/51007 on Notification 2003/475/A; SG(2003) D/52110 on Notification 20-03/200/A; SG(2005) D/50052 on Notification 2004/311/A. COM (2006) 104 final, above n. 7, discusses a number of notifications.

31 The final quotation is from European Commission, Detailed Opinion under Directive 1998/34/EC Laying Down a Procedure for the Provision of Information in the Field of Technical Standards and Regulations [1998] OJ L 204/37 *Notification 2003/475/A (Genetic Engineering Precautionary Measures Act).*

32 See the discussion in Les Levidow and Karin Boschert, 'Coexistence or Contradiction: GM Crops versus Alternative Agricultures in Europe' (2008) *Geoforum* 174. For detail on GM-free regions, see http://genet.iskra.net/ (accessed December 2007). A number of EU regions made commitments to GM-free agriculture in *Charter of the Regions and Local Authorities of Europe on the Subject of Coexistence of Genetically Modified Crops with Traditional and Organic Farming* (Florence, 2005).

is neither irrational nor a purely economic issue. At the very least, it might be argued that maintaining some 'GM-free' production is a precautionary measure allowing a response to any unforeseen health or environmental problems. We should also note that identifying adventitious GM material in products is a genuine and ongoing challenge as new modifications are developed, again emphasising the value of prevention.[33] And in Member States with small fields and small farms, certain coexistence measures like isolation distances between crops become very difficult – a GM-free zone lessens the practical difficulties of coexistence.

The legality of GM-free zones, however, is at least debatable. The Commission's guidelines provide that coexistence should be carried out on the 'appropriate scale', prioritising 'farm-specific management measures and . . . measures aimed at coordination between neighbouring farms', and also particular crop species, rather than all GMOs.[34] Indeed, the whole tenor of the Recommendation is against GM-free zones, even on a GMO-by-GMO basis.[35] In the *Austrian* case discussed in Chapter 3,[36] the Commission stated that to be proportionate (and disproportionate measures would generally be unlawful), measures must would 'take account of' three factors: 'specific crop type'; 'specific crop use'; and 'if sufficient levels of purity cannot be achieved by other means'. Advocate General Sharpston confirmed when the Austrian decision reached the ECJ that whilst a range of measures remain open, a Member State may not 'adopt legislation imposing a blanket ban on GMOs within their territory, unless they can provide evidence which complies with all the criteria in Article 95(5) EC'.[37]

---

[33] European Commission, Staff Working Document, *Second Report on the Experience of Member States with GMOs Placed on the Market under Directive 2001/18/EC on the Deliberate Release into the Environment of Genetically Modified Organisms* SEC (2007) 274, para. 32; European Commission, Staff Working Document, *Communication on the Mid-term Review of the Strategy on Life Sciences and Biotechnology* SEC (2007) 441, p. 45.

[34] Commission Recommendation, above n. 3, paras 2.1.5 and 2.1.6.

[35] See also COM (2006) 104 final, above n. 7, p. 5.

[36] Commission Decision 2003/653 Relating to National Provisions on Banning the Use of Genetically Modified Organisms in the Region of Upper Austria notified by the Republic of Austria pursuant to Article 95(5) of the EC Treaty [2003] OJ L 230/34. See T-366/03 and 235/04 *Land Oberosterreich and Austria v Commission* [2005] ECR II-4005; Case C-439/05, above n. 16.

[37] *Land Oberosterreich*, above n. 16, para. 147. They may, however, 'pass legislation such as that finally adopted in Upper Austria, which imposes stringent safeguards on the cultivation of GMOs. They may intervene in the Community approval process under part C of Directive 2001/18. They must monitor and report new information under Article 20 of that directive, they may invoke the safeguard clause in Article 23 and they may take measures under Article 26a. Moreover, in accordance

The Commission has, however, long been aware of the political sensitivity of GM-free zones,[38] and has conceded that regional measures 'could be considered', when 'sufficient levels of purity cannot be achieved by other means', but on as limited a geographical scale as possible, and subject to justification for each crop and product type.[39] This is a significant concession, although 'sufficient' levels of purity are presumably defined by reference to legislative thresholds for acceptable GM presence in non-GM products. This is discussed in detail below, but for current purposes it suffices to say that a product does not have to be labelled as containing GMOs if it unintentionally contains up to 0.9 per cent GM. Defining this as a 'sufficient' level of purity is highly contentious, as will be explored in the next section. More recently, the Commission has said that GM-free initiatives do not require notification by the Member States provided they are: 'a mere declaration of intent, a description of the status quo, or are based on voluntary agreements of all stakeholders concerned and do not imply a prohibition of the use of authorised products'.[40] The weakness of the concession is obvious, requiring the commitment of everybody involved and really no formal 'regulation'. The Commission has even gone so far as to demand 'the written consent of the persons cultivating and owning this land' before voluntary measures are taken, because 'the establishment of areas free from genetically modified varieties could have an impact on the value of cultivable land'.[41] Much of the burden of ensuring (entirely or as close to entirely as possible) GM-free produce seems to fall on voluntary agreements, or spontaneous arrangements between farmers.[42] Quite how effective that will be is open to question, and clearly any dissenting landowner can scupper the plan.

---

with the 2003 guidelines, they may and should take steps to ensure that any cultivation of GMOs within their territory can coexist, without admixture, with neighbouring conventional or organic farming', para. 146.

38 'From a political point of view it could be difficult to reject these attempts at establishing GM-free zones, which are driven by strong public local concern and economic considerations (such as protection of local traditional agriculture), without offering some alternative solutions together with the necessary legal clarity and guidance to address their concerns and considerations', *Communication to the Commission for an Orientation Debate on Genetically Modified Organisms and Related Issues* (2004), p. 5.

39 Commission Recommendation, above n. 3, para. 2.1.5.

40 SEC (2006) 313, above n. 13, p. 8.

41 European Commission, Detailed Opinion under Directive 1998/34/EC Laying Down a Procedure for the Provision of Information in the Field of Technical Standards and Regulations [1998] OJ L 204/37, *Notification 2006/73/P (Draft order regulating the conditions and procedure for establishing areas free from the cultivation of genetically modified varieties).*

42 But there are also legal questions. First of all, there must be a question around any Member State instigation of, and definitely participation in, voluntary agreements

The Commission also anticipates that its Coexistence Bureau will make recommendations for regions where farming conditions make farm-level coexistence difficult to achieve, and it will be interesting to see whether this goes beyond voluntary and unprompted measures and how influential they are in the Member States.[43] In many areas of EU social regulation, the more complex approaches to 'multi-level governance' involve a 'network' somewhere. Coexistence is no exception. As well as the Bureau, composed of committees of experts working on more technical crop-specific guidance documents, the Commission has set up a 'coexistence network',[44] both to 'facilitate the exchange of information about ongoing and planned research projects at national and Community level',[45] and to 'gather and coordinate information based on studies at Community and national level, and to observe the development regarding coexistence in the Member States'.[46] The networking group consists of national experts appointed by the Member States, chaired by a Commission representative, and subject to the invitation (by the Commission representative) of 'other experts'. Chapter 3 touched on the subject of 'networks'. In the authorisation context discussed there, networks are called on to mitigate the centralisation inherent in the role of the European Food Safety Authority. Networking of central and national experts is thought to provide some compromise between the efficiency of central action and national democratic influence. In the context of coexistence, this works in reverse, in that the network prevents the surrendering of central and mutual (that is, Member State on Member State) influence in a context of apparent national independence.

The role of the network is innocuous enough, information based, focusing 'on results of scientific studies as well as best practices'.[47] Networks are an important part of multi-level governance, in theory avoiding any formal and

---

from an internal market perspective. Further, depending on the particular agreements reached, competition law issues might be raised by agreements not to grow particular crops. The Commission has at least shown itself willing to make an effort to take into account environmental considerations in competition policy. However, recall that the Commission insists that coexistence is not an *environmental* but an economic issue.

[43] European Commission, Staff Working Document, *Accompanying the Communication on the Mid-term Review of the Strategy on Life Sciences and Biotechnology* SEC (2007) 441, p. 42.

[44] Commission Decision 2005/463/EC *Establishing a Network of Groups for the Exchange and Coordination of Information Concerning Coexistence of Genetically Modified, Conventional and Organic Crops* [2005] OJ L 164/50.

[45] Decision 2005/463, above n. 44, Recital 1, referring to Commission Recommendation, above n. 3.

[46] Decision 2005/463, above n. 44, Recital 2.

[47] Commission Decision 2005/463, above n. 44, Recital 3. See also SEC (2006) 313, above n. 13, p. 21.

hierarchical relationship between the centre and the Member State, and not readily defined as national or EU in their operation. Networking may, however, turn out to be one more constraint on the national flexibility apparent on the face of legislation. It could result in a form of 'soft harmonisation', with centralised expectations (if not binding standards) established by the network rather than in legislation. The Integrated Pollution Prevention and Control Directive is a famous case of increased local flexibility in implementation, mainly through the open-ended nature of the primary standard in the legislation, that of 'best available techniques' (BAT).[48] However, 'BAT reference notes' (or BREFs), written by committees of Member State representatives, industry and environmental interest groups, set out BAT for particular sectors or issues. BREFs are not formally binding, and should simply be a factor to take into account when determining BAT. They are, though, obviously a tempting reference point for regulators, industry and environmental interest groups. The output of our coexistence networks could also carry considerable scientific and political, if not legal, authority. Whilst the centre is not issuing any command, these tools of 'governance' could turn out to be an effective way of further emphasising the central hold on coexistence initiated by the Commission's Recommendation. A non-hierarchical, soft form of 'Europeanisation' could have very positive implications, preventing the national shirking of environmental or social responsibilities, and the absence of central coexistence obligations is after all a major gap in an otherwise intensively harmonised area of regulation. Moreover, the Commission seems to be taking a very narrow approach to coexistence at the moment, and the networks could exercise a positive influence on the Commission's perspective. Even at this progressive best, however, networks tend to lack transparency whilst being most profoundly political in their impact.

To say that coexistence is a national competence is clearly only a very small part of the picture. We have a complex set of arrangements, characterised perhaps less by multi-level governance than by a battle for authority. The Commission resisted harmonisation of coexistence, but efforts of the Member States to fill the resulting gap are in turn resisted. The Commission is trying to wrest back authority on this issue (so far, it must be said, successfully), in

---

48 Directive 1996/61/EC on Integrated Pollution Prevention and Control [1996] OJ L 257/26. See the discussion in Maria Lee, *EU Environmental Law: Challenges, Change and Decision-Making* (Hart Publishing, 2005), Chapter 6. Note also the importance of networks in the implementation of the Water Framework Directive (Directive 2000/60/EC establishing a Framework for Community Action in the Field of Water Policy [2000] OJ L 327/1); see Joanne Scott and Jane Holder, 'Law and New Environmental Governance in the European Union' in Gráinne de Búrca and Joanne Scott (eds), *Law and New Governance in the EU and the US* (Hart Publishing, 2006).

particular by its insistence that coexistence is a purely economic issue. This in turn emphasises the most restrictive aspects of basic internal market law, over which the Commission strives to maintain control in its application of the Directive 1998/34 process.

Coexistence and Thresholds of 'Contamination'

Defining coexistence as a solely economic issue emphasises the importance of the legislative thresholds for mandatory labelling of GMOs and for seed purity. This in turn further restricts national autonomy on coexistence, because, as we shall see, a regulatory policy of purity (or something approaching purity) is simply not envisaged by the Commission. Purity is irrelevant for as long as impurity has no economic impact, that is, as long as the market accepts impurity: and the market is presumed not to discriminate (because it cannot) between unlabelled products.[49]

All products (food and feed or otherwise) containing or consisting of GMOs, plus food and feed products produced from GMOs, must be labelled as such. *Any* intentional or technically avoidable presence of GM material, however small, would need to be labelled. Labelling is not, however, required if the product, or an individual ingredient in the product, contains up to 0.9 per cent (authorised) GM material,[50] provided that the presence of GM material is 'adventitious or technically unavoidable'.[51] Setting the threshold is highly controversial,[52] but so is the use of the threshold. To take advantage of the 0.9

---

49 There may be voluntary 'GM-free' labelling, which could include the imposition of higher standards in the organic sector, but the voluntary approach is likely to be very challenging, as discussed further below.

50 Thresholds have been set for food and feed, but not for seeds. Trying to set a threshold of GM presence in seeds is proving extraordinarily controversial, with disagreement within the Commission as well as between Member States and institutions. Thresholds between 0.3 and 0.7 per cent have been proposed by the Commission for different seeds; see the discussion in Schenkelaars Biotechnology Consultancy, with Risk and Policy Analysts Ltd, *Means to Improve the Consistency and Efficiency of the Legislative Framework in the Field of Biotechnology Article 31 of Directive 2001/18/EC* (Report for the Commission, 2004), pp. 127–33.

51 Regulation 1829/2003, above n. 12, Article 12(2); Regulation 1830/2003 concerning the Traceability and Labelling of Genetically Modified Organisms and the Traceability of Food and Feed Products Produced from Genetically Modified Organisms and amending Directive 2001/18/EC [2003] OJ L 268/24, Article 4(6) and (7) – note that there is no 'appropriate steps' proviso in this Regulation.

52 The level of 0.9 per cent can be amended by comitology to reflect scientific and technical advances. Under European Commission, *Proposal for a Directive amending Directive 2001/18 concerning the deliberate release into the environment of genetically modified organisms, as regards the implementing powers conferred to the*

per cent 'adventitious' or 'technically unavoidable' exception to labelling obligations, evidence must be provided to establish that 'appropriate steps' have been taken to avoid the presence of GM material. These three phrases are clearly open to different interpretations, but they at least suggest that labelling obligations are not avoidable routinely, but only where certain efforts have been taken to avoid GM. The wording may even suggest that, to avoid labelling obligations, the efforts of the producer must be to avoid GMOs entirely or as much as possible, with the threshold providing an exceptional and pragmatic response to the pervasiveness of living organisms. This interpretation has, however, been rejected by the Commission. The Commission's approach to the coexistence measures allowed to the Member States is defined by reference to 'legal obligations for labelling and/or purity standards',[53] that is, the 0.9 per cent threshold. The Commission assumes that thresholds set in the legislation are entirely unproblematic, transforming the labelling thresholds into a threshold for regulatory intervention on coexistence. The Commission suggests that it would be illegitimate for a Member State to apply mandatory regulatory measures aiming below that threshold:

> Measures for co-existence should be efficient and cost-effective, and proportionate. They shall not go beyond what is necessary in order to ensure that adventitious traces of GMOs stay below the tolerance thresholds set out in Community legislation.[54]

As mentioned above, those who seek to rely on the 0.9 per cent threshold are obliged to establish that 'they have taken appropriate steps to avoid the presence of such material'.[55] Meeting a national regulatory authority's instructions on coexistence may well constitute 'appropriate steps'. In this case, the appropriate steps to avoid contamination are assessed by reference to coexistence requirements – and they, according to the Commission, should themselves be defined by reference to the thresholds. This deprives the requirement to take appropriate steps of much of its restrictive value, turning it into an effort to

---

Commission COM (2006) 920 final, this will be subject to European Parliament involvement through the new 'regulatory procedure with scrutiny'; see Chapter 3, n. 8. Department for Environment, Food and Rural Affairs, *Consultation on Proposals for Managing the Coexistence of GM, Conventional and Organic Crops* (DEFRA, 2006), discusses the ways in which limitations in the technology restrict the options. See also Commission Recommendation 2004/787/EC on technical guidance for sampling and detection of genetically modified organisms and material produced from genetically modified organisms as or in products in the context of Regulation (EC) 1830/2003.

53 Commission Recommendation, above n. 3, Recital 3. This is confirmed most emphatically throughout the Commission's responses to 1998/34 notifications.

54 Commission Recommendation, above n. 3, para. 2.1.4.

55 Regulation 1829/2003, above n. 12, Article 12(3).

keep contamination below the 0.9 per cent threshold rather than to avoid unintentional contamination.[56] The Commission's response to Hungary's proposed coexistence measures reinforces this. Hungary would have required crops produced in a 'refuge zone'[57] to be labelled as 'plants modified by gene technology', regardless of how much GM material is present. The Commission insists that labelling is not required of 'products containing traces of GMOs which do not exceed 0.9 per cent provided that these traces are adventitious and technically unavoidable'.[58] It is at least arguable that the fact of growing in this buffer zone means that 'appropriate steps' have *not* been taken to avoid adventitious presence, albeit that the level of GM presence is below 0.9 per cent. This may need further analysis on the facts, but is simply not pursued by the Commission. Aiming for zero but taking a pragmatic approach to the reality of falling short on occasion is a very different philosophy from *aiming* for the 0.9 per cent threshold. The legislation does not support the Commission's view that the latter is appropriate.

If the Commission's interpretation of 'appropriate steps' empties it of any obligation to aim to avoid contamination, the 'appropriate steps' obligation may nevertheless provide a sort of penalty (at least in the form of an obligation to label) for failure to comply with coexistence obligations. This is no mean feat. However, the 'appropriate steps' obligation is generally going to be the responsibility of a different party from that responsible for the coexistence regulations imposed by the Member State. 'Appropriate steps' apply to the farmer (or retailer) aiming to avoid GMOs (and associated labelling obligations). But, under the Recommendation, coexistence measures are the responsibility of the farmer introducing 'a new product type' – at least in the early stages, the GM farmer. The more closely one looks at the legal context for coexistence (including the liability and intellectual property regimes discussed below and in Chapter 5), the more apparent it becomes that the burden of coexistence will fall on farmers seeking to avoid GMOs.

The 0.9 per cent threshold is a particular concern for the organic sector, where 'GM-free' status can be especially important. If GM agriculture becomes widespread across a range of agricultural products, it will be difficult for organic farmers to aim below 0.9 per cent GM presence without regulatory

56 See the discussion in Lee, above n. 48.

57 A refuge zone is the area within the buffer zone surrounding a cultivation of GM plants cultivated with plants of identical species which are not genetically modified.

58 European Commission, Detailed Opinion under Directive 1998/34/EC Laying Down a Procedure for the Provision of Information in the Field of Technical Standards and Regulations [1998] OJ L 204/37, *Notification 2005/634-637/HU (Act on the amendment and modification of the law on gene technology activity).*

assistance. The UK provides a good example of the complexities of the debate.[59] Many organic certifiers, including the Soil Association, the UK's largest organic certifier, have a GM-free policy, which effectively means that only 0.1 per cent (surrogate zero, the lowest detectable level) GM presence is permissible in products labelled organic. The UK Government, however, clearly thinks that the 'GM-free' option 'would present serious difficulties and ultimately may not be in the best interests of the organic sector', essentially because it will be unachievable: 'It could undermine consumer confidence in the integrity of organic produce if there were repeated breaches of a specific GM threshold and/or it had subsequently to be abandoned as impractical'.[60]

The legal position is that GMOs cannot be *used* in organic agriculture.[61] The *presence* of GMOs is more complicated. A new Regulation on organic farming was agreed in 2007, confirming explicitly that the 0.9 per cent threshold also applies to organic produce, meaning that a product that has to be labelled under the GMO legislation cannot also be labelled organic.[62] Organic certification bodies can voluntarily require their members to go beyond the minimum regulatory standards, or organic farmers and producers could even voluntarily aim individually for zero contamination; nor is the voluntary labelling of GM presence between 0.1 per cent and 0.9 per cent prohibited. However, the burden for achieving these more stringent standards is by this Regulation passed to the organic sector. Whether it will be practical to avoid GM presence in the absence of regulatory assistance remains to be seen.[63] The recitals to the Organic Regulation place a different emphasis on the thresholds than does the Commission's GMO policy, especially in their emphasis on the effort that needs to be taken to avoid GM presence:

---

59 I use the UK as an example in this chapter, but of course other Member States take different approaches. So Austria and Germany, for example, are standing firm on the purity of organic produce.

60 DEFRA, above n. 52, paras 121 and 122. This counsel of doom may encourage supporters of organic production to conclude that, rather than allowing GMOs to change the nature of organic produce, GM agriculture should not be allowed at all in the UK. The government attempts to pre-empt this by claiming that it would anyway be impossible to keep organic produce 'GM-free' because of global agricultural imports, para. 121.

61 Council Regulation 834/2007/EC on Organic Production and Labelling of Organic Products and Repealing Regulation (EEC) 2092/91 [2007] OJ L 189/1, Article 9.

62 Regulation 834/2007, above n. 61, Article 23(3).

63 UK Government, above n. 52, consulted on a lower threshold than 0.9 per cent for organics ('say 0.5 per cent'), para. 109, although this predates the agreement of Regulation 834/2007, above n. 61.

> The aim is to have the lowest possible presence of GMOs in organic products. The existing labelling thresholds represent ceilings which are exclusively linked to the adventitious and technically unavoidable presence of GMOs.[64]

This is not, however, pursued beyond the recitals to the Regulation. Moreover the Commission 'stresses' proportionality when discussing national proposals to apply different coexistence measures depending on whether the non-GM crops are conventional or organic 'even though the same labelling thresholds . . . apply'.[65] The Commission assumes generally that any effort by Member States to aim below 0.9 per cent adventitious presence goes beyond the scope of lawful measures, and it makes no special allowance for organic agriculture.[66]

We are seeing a subtle, and centralised, redefinition of 'organic' in order to allow for a particular understanding of coexistence. It is perfectly arguable that a lower threshold would be appropriate for organic agriculture, so that organic does not mean 'non-GM' in the sense of the legislation, but gets closer to 'GM-free'. Measures for the coexistence of GM and organic farming are in part about allowing organic farming to continue to thrive (as is apparent from the UK government document cited above – which also indicates that there is real disagreement about what thriving requires). The Commission elsewhere accepts that organic farming is a form of farming 'known to deliver public goods', including 'public health, social and rural development' and 'animal welfare' as well as environmental benefits.[67] According to the Organic Regulation, organic production has a 'dual societal role':

> where it on the one hand provides for a specific market responding to a consumer demand for organic products, and on the other hand delivers public goods contributing to the protection of the environment and animal welfare, as well as to rural development.[68]

---

64 Regulation 834/2007, above n. 61, Recital 10.

65 COM (2006) 104 final, above n. 7, p. 6. Note that this was the Commission's position even before Regulation 834/2007 was agreed; see the discussion in Lee, above n. 48, pp. 255–9.

66 This is also a consistent position in the Detailed Opinions under Directive 1998/34; see especially 2005/610/A, above n. 10, para. 2(a).

67 European Commission, *European Action Plan for Organic Food and Farming* COM (2004) 415 final, section 1.4.

68 Regulation 834/2007, above n. 61, Recital 1. Note also that organic products are one of the 'sustainable' products promoted in the EU's *Renewed Sustainable Development Strategy* (2006), p. 13. A very broad range of goals is associated with sustainable development in this document.

On this basis, the long-term credibility of organic farming has public benefits. In particular, its distinctiveness from more industrialised forms of farming, and associated thriving in the market place, is also an issue of broader public goods, not simply an economic issue. In banning the *use* of GMOs in organic production, it is acknowledged that they are 'incompatible with the concept of organic production and consumers' perception of organic products'.[69] An organic label does mean 'GM-free' in that the bar on the use of GMOs means that organic animal products have not been reared with GM feed, and also applies to processed non-food (for example, organic cotton). But, in terms of GM content, it will be difficult for the organic sector to aim below the 0.9 per cent threshold given the regulatory position on the issue. Yet again, the compromise reached in the interests of allowing commercial exploitation of GMOs is distinctly uncomfortable.

Conclusions: Coexistence as an Economic Problem?

Conceptualising coexistence as always and only an economic issue has severely restrictive legal consequences. If coexistence were more than an economic issue, the Member States would of course still be constrained by EU law; in particular, action would still need to be proportionate, scientific justification would be demanded and Commission and judicial scrutiny would be intense. But there would be space in which genuinely competing public policy arguments could be heard. The openness of decision making is limited by an economic perspective on coexistence, excluding substantive discussion of non-economic coexistence issues. In addition, the lay public and environmental interest groups may not so easily anticipate their very important role in this debate. The Commission's Recommendation, rather unusually albeit probably not deliberately restrictively, encourages 'stakeholder' rather than 'public' involvement in decisions on coexistence.[70] It is more difficult for outsiders to argue that they have a 'stake' in economic decision making, and whilst organic (and conventional) farmers have a clear economic 'stake' in coexistence, this reduces their contribution to one of narrow self-interest, on a par with the economic interests of the biotechnology industry.

The *Austria* case discussed in Chapter 3 is again illustrative. The Austrian measures were designed primarily to protect organic agriculture, although they were claimed by Austria also to protect conventional agriculture and biodiversity. Austria considers its agriculture to be characterised by small structured farming (small farms, small fields) with high levels of organic farming (about

69 Regulation 834/2007, above n. 61, Recital 9.
70 Commission Recommendation, above n. 3, para. 2.1.2.

7 per cent in Upper Austria). Most pertinently for current purposes, the Commission found that the Austrian measures did not 'specifically [concern] the protection of the environment or the working environment',[71] but that it 'appears that Austrian concerns about coexistence relate more to a socio-economic problem than to the protection of the environment or the working environment'.[72] Whatever the administrative convenience, drawing a clear line between public goods and economic concerns is not easy. The public goods provided by farming, and especially organic farming, are also 'at the heart of policies on environmental protection, food security and consumer concerns (health and food safety, animal welfare issues and the viability of rural areas)'.[73] The Commission and other EU institutions have recently been paying greater attention to agriculture's 'multi-functional role in society': as well as food production, agriculture contributes to the 'sustainable development of rural areas' and environmental and biodiversity protection.[74] These public goods are particularly pertinent in the case of organic farming, in which context the purely economic perspective on coexistence is especially difficult.

There are also more specific environmental issues in the coexistence agenda. So, for example, changes to pest and weed management in non-GM farming, because of GM presence in a crop, could have longer-term environmental impacts, as well as economic impacts. This matter should be addressed in the risk assessment pre-authorisation, taking us back to the residual (economic) nature of coexistence measures. There is indeed a 'no adverse effects' approach in the regulatory framework. Which brings us to a fairly fundamental issue. Seeing coexistence as an economic issue crucially assumes the adequacy of the authorisation process. And particularly when we think about the profound changes that widespread commercial use of agricultural biotechnology *could* instigate, and potentially highly localised impacts in the

---

71 Commission Decision 2003/653, above n. 36, para. 66.

72 Commission Decision 2003/653, above n. 36, para. 67. The Committee of the Regions emphasises the 'intrinsic rather than economic value' of different forms of agriculture, *Opinion of the Committee of the Regions on the Communication from the Commission to the Council and the European Parliament Report on the Implementation of National Measures on the Coexistence of Genetically Modified Crops with Conventional and Organic Farming* [2007] OJ C 57/11.

73 Mary Footer, 'A Tale of Two Commons: Plant Genetic Resources and Agricultural Trade Reform' in Somsen, above n. 18. What organic farming means to the consumer is complex. Gill Seyfang, 'Cultivating Carrots and Community: Local Organic Food and Sustainable Consumption' (2007) 16 *Environmental Values* 105, identifies three, sometimes competing, paradigms for the local organic product: 'as a tool for creating green localised economies, as health-conscious global food for supermarket shoppers, and as reactionary fare for status-driven traditionalists', p. 119.

74 European Commission, *Agriculture and Environment* (2003). See for detailed discussion http://ec.europa.eu/agriculture/envir/index_en.htm.

different ecological and agricultural contexts around the EU, even a very forward thinking system of risk regulation ('direct or indirect, immediate or delayed' including 'cumulative long-term effects'[75]) could easily fall short. The EU level authorisation applies in principle across the whole of the EU, notwithstanding the mighty challenge of incorporating all ecological, agricultural and social conditions in that single authorisation.[76]

The Commission continues to insist that coexistence measures cannot be used to restrict cultivation of GMOs in ecologically sensitive areas,[77] assuming that the risk assessment dealt adequately with all environmental concerns. Whilst there can be no blanket ban, however, the Commission does accept that the Habitats Directive might be applied to the cultivation of GMOs.[78] This Directive at least allows, in fact requires, a focused and location-specific consideration of the ecological impacts of GMOs on sites designated as part of the Natura 2000 network. The Habitats Directive applies an authorisation procedure to any 'plan or project' that is 'likely to have a significant effect' on a site. The plan or project must undergo 'appropriate assessment of its implications for the site in view of the site's conservation objectives', and if the 'integrity of the site' would be adversely affected, consent must *prima facie* be refused.[79] The pivotal concept of the 'integrity of the site' is undefined in the legislation, but interpreted by the Commission as relating to the coherence of the site's ecological structure and function, which enables it to sustain the habitat and the levels of populations of species for which it was

75 Directive 2001/18, above n. 1, Article 2(8) and Annex II.

76 Because GMOs are products, they are a rather different policy area from other environmental legislation, where, although more or less flexible standards (of emissions or of the quality of the receiving environment) may be set centrally, individual facilities apply for authorisation locally. Remember also that some GMOs were authorised before the new Member States had joined the EU and were able to contribute to the setting of conditions under Directive 2001/18, above n. 1, Article 19. See also 2785th Council meeting, 20 February 2007, in respect of Hungary's measures on MON810 maize: 'the different agricultural structures and regional ecological characteristics in the European Union need to be taken into account in a more systematic manner in the environmental risk assessment of GMOs'. See Chapter 3, p. 90.

77 COM (2006) 104 final, above n. 7, p. 5.

78 See the Detailed Opinions, above n. 30, and SEC (2006) 313, above n. 13. Directive 1992/43 on the conservation of natural habitats and of wild fauna and flora [1992] OJ L 206/7. Also Directive 1979/409 on the conservation of wild birds [1979] OJ L 103/1.

79 Article 6. The project may nevertheless go ahead, subject to the provision of 'compensatory measures', for 'imperative reasons of overriding public interest, including those of a social or economic nature'; if the site hosts a priority habitat or a priority species, unless the Commission gives a positive opinion, it can only go ahead for reasons relating to human health, public safety or environmental protection.

classified.[80] The potential to apply nature conservation legislation to the cultivation of even authorised GMOs is a major comfort, especially given the ECJ's highly precautionary interpretation of the Habitats Directive.[81] It is likely, though, only occasionally to inhibit the release of GMOs, as scientific evidence of specific potential harm will be demanded, and in any event only in respect of protected sites (although according to the Commission this amounts to 20 per cent of the EU[82]). But, as well as providing protection *in extremis*, the continued relevance of nature conservation law is also an acknowledgement that some environmental concerns survive authorisation.

And finally, an economic perspective falls short because ethical and political concerns are formally (albeit not entirely) discounted in the authorisation process. The pattern of the regulation is that these concerns should be swept up by consumer choice, by the obligation to label GMOs. Meaningful coexistence is absolutely central to the consumer choice rhetoric surrounding the EU regulation of GMOs, since in the absence of distinctiveness between different forms of farming consumers cannot choose. Consumer choice at the very least should be a relevant consideration in coexistence measures, taking us beyond the Commission's solely economic approach. Even if we were to accept that coexistence has nothing to do with the environment or human health, it is nevertheless intimately concerned with the 'other issues' legitimately provoked by agricultural biotechnology.

The Commission is not of course the ultimate interpreter of the Treaty or the legislation on GMOs. That task remains with the ECJ, and the Commission's Recommendation must comply with the Treaty and regulatory framework. The understanding of coexistence as purely an economic issue rests on a flawed approach to the role of the law after authorisation of a GMO. Whilst there are good grounds for challenging the Commission's approach, however, the legal background is well entrenched. We should also note that Advocate General Sharpston considers that the Commission 'correctly sought to distinguish between the environmental issues . . . and the socio-economic issues of agricultural management'.[83] A more modest task would be to examine the Commission's approach to thresholds in the legislation, which is thoroughly misconceived. Rather than constituting a norm, 0.9 per cent should

---

[80] European Commission, *Managing Natura 2000 Sites* (European Communities, 2000), para. 4.6.3.

[81] Case C-127/02 *Landelijke Vereniging tot Behoud van de Waddenzee v Staatssecretaris van Landbouw Natuurbeheer en Visserij* (*Cockle Fishers*) [2004] ECR I-7405.

[82] See the Commission website: http://ec.europa.eu/environment/nature/index_en.htm.

[83] *Land Oberosterreich*, above n. 16, para. 118.

constitute a wholly exceptional threshold,[84] a pragmatic response to fallibility rather than an aspiration. On this basis it would be legitimate for Member States to introduce measures aiming below 0.9 per cent, and if this level of purity can only be achieved by regional GM-free zones, so be it. It is not clear why a redefinition of what we mean by 'organic' and 'GM-free' is a positive conclusion to the debate on coexistence. It is the only practical response to the coexistence of agricultural biotechnology only if it is assumed that widespread GM agriculture is an objective and unavoidable fact, rather than a choice.

Coexistence is a subtle problem, and requires a far more subtle solution than has so far been offered by the Commission. We have a double bind, as the Commission declines responsibility for ensuring coexistence but seriously restricts national efforts to bridge that important gap. The approach of the Commission is presented as a 'neutral' solution to resistance to agricultural biotechnology, on the assumption that it allows farmers and consumers as individuals to reject GMOs. But it is far from neutral in practice. Compromise is very difficult, and there are many situations in which different forms of agriculture will be in direct conflict, and not only symbolically. The flourishing of non-GM and organic agriculture is in part an issue of public goods. These public goods cannot even be heard if coexistence is understood as a wholly economic issue.

CIVIL LIABILITY FOR NEGATIVE IMPACTS OF GMOs

One of the matters consistently raised during the EU moratorium on new authorisations of GM crops was liability for any adverse consequences of GMOs. And indeed the Deliberate Release Directive refers explicitly to liability. Not only is the Directive declared to be without prejudice to national 'legislation' in 'the field of environmental liability', but also 'Community legislation in this field needs to be complemented by rules covering liability for different types of environmental damage in all areas of the European Union'.[85] This formulation represents a compromise for those legislators who wanted to see strict liability introduced in the Deliberate Release Directive itself, and seems to accept at least that a specific liability regime is required in respect of possible *environmental* damage.

---

84 See also the advice provided to Friends of the Earth, Paul Lasok and Rebecca Haynes, 'In the Matter of DEFRA's Proposals for Managing the Coexistence of GM, Conventional and Organic Crops' (2006).

85 Directive 2001/18, above n. 1, Recital 16. Presumably the reference to 'legislation' includes common law (judge-made rather than statute).

The Community legislation anticipated in the Deliberate Release Directive is the Environmental Liability Directive, agreed after years of wrangling in 2004.[86] Perhaps wisely, given the complexity and controversy of any such project, the Directive steps back from earlier proposals for major changes to national tort law in an environmental context[87] and 'liability' is arguably even something of a misnomer. The Directive instead requires Member States to set up a purely administrative scheme in which regulators require the restoration or prevention of 'environmental damage' by operators.

The debate about liability extends beyond the narrow scope of the Environmental Liability Directive, discussed further below. So, for example, non-GM farmers may lose market value through having to label their conventional or organic produce as containing GMOs, or may have to change their agricultural management (where to plant, pesticide and herbicide regime, dealing with volunteers, and so on[88]). There may also be cases of illness or injury as a result of GM food or crops. Liability arrangements are primarily a Member State responsibility, and this is not the place for a detailed examination of the application of tort law to the range of potential harms, if only because the detail is highly jurisdiction specific. The English law will be used as an example as the need arises, although other jurisdictions are taking interesting legislative approaches.[89] Instead of trying to apply tort law to GMOs,

---

86 Directive 2004/35 on Environmental Liability with Regard to the Prevention and Remedying of Environmental Damage [2004] OJ L 143/56. Interest in 'environmental liability' began with European Commission, *Proposal for a Council Directive on Civil Liability for Damage Caused by Waste* [1989] OJ C 251/3, which never became legislation.

87 European Commission, *White Paper on Environmental Liability* COM (2000) 66 final. On fundamental changes in approach in the different documents leading up to more general liability, see Maria Lee, 'The Changing Aims of Environmental Liability' (2002) 11 *Environmental Law and Management* 185.

88 Note that Percy Schmeiser, the farmer sued by Monsanto for patent infringement in the Canadian litigation discussed in Chapter 4, is in another legal dispute with Monsanto. Monsanto's patented canola appeared in his fields again. He called Monsanto to remove the patented material, but refused to sign their waiver. He therefore had to remove the material himself, and sent Monsanto a bill, which they have not paid. See Sean Pratt, 'Schmeiser Renews Monsanto Battle' *Western Producer* 31 May 2007.

89 For detail on potential liability in the English common law and in European legislation, see Maria Lee and Robert Burrell, 'Liability for the Escape of GM Seeds: Pursuing the "Victim"?' (2002) 65 *Modern Law Review* 517; Christopher Rodgers, 'Liability for the Release of GMOs into the Environment: Exploring the Boundaries of Nuisance' (2002) 62 *Cambridge Law Journal* 371; Maria Lee, 'Regulatory Solutions for GMOs in Europe: The Problem of Liability' (2003) 13 *Journal of Environmental Law and Practice* 311. For discussion of some legislative proposals in this area, see SEC (2006) 313, above n. 13.

some of the particular challenges presented by agricultural biotechnology are examined here. Just as large-scale, sometimes temporally and physically remote mechanised accidents challenged tort law as a 'subject' from the 19th century, so can new technologies in the 21st. Specifically, liability is explored here along three key parameters: the type of harm, the identification of the parties to litigation and the nature of liability. Where the responsibility for harms brought about by GMOs does and should lie is proving to be a very difficult legal question. This is a regulatory issue, distributing risks, costs and benefits and influencing behaviour.

The allocation of liability presents challenges both in terms of predictability and in terms of the suitability of the likely outcome. Predictability will emerge only slowly and piecemeal if liability is assessed by the courts on existing principles, as cases arise. Certainly the English common law is very far from any clear and predictable liability scheme, and international debate on the subject under the framework of the Cartagena Protocol on Biosafety remains profoundly open on the most significant questions.[90] Whether there will be any transfer of costs through tort liability remains uncertain, and actual and proposed legislative intervention makes only partial inroads into that uncertainty.[91] Some uncertainty is inherent in any liability provisions, but given the painfully slow and exceptionally controversial negotiation of the authorisation and coexistence regulation in this area, combined with the inherent uncertainty around the impact of GMOs, predictable liability arrangements should be an important element of the overall regulatory framework. Whilst flexibility is an important value to place alongside predictability, who bears the risk of ill effect is a deeply political question. It is inextricably linked with the overall regulatory regime, and should have been addressed within that context. And if predictable liability rules allow assessment of the overall regulatory scheme applying to GMOs, the assessment reached will depend on the allocation of liability, which in practice is the controversial and stimulating question.

---

[90] The Cartagena Protocol is discussed in Chapter 6. See *Report of the Open-Ended Ad Hoc Working Group of Legal and Technical Experts on Liability and Redress in the Context of the Cartagena Protocol on Biosafety on the Work of its Fourth Meeting* UNEP/CBC/BS/WG-L&R/4/3, 13 November 2007, which puts forward a 'blueprint' containing a range of options on, for example, the meaning of damage and the nature of liability.

[91] Actual legislative intervention can be found in Directive 2004/35, above n. 86; Directive 1985/374/EEC on the Approximation of the Laws, Regulations and Administrative Provisions of the Member States concerning Liability for Defective Products [1985] OJ L 210/29, amended by Directive 1999/34/EC [1999] OJ L 141/20. A number of Member States have either introduced or are considering liability schemes. Note that the Commission is studying liability, SEC (2007) 441, above n. 33, p. 40.

The Nature of Harm

The first challenge for liability regimes faced with negative impacts of GM crops will be fitting harm within a traditional tortious framework. The main types of potential harm are, first, illness or disease caused by consumption of GMOs or exposure to a GM pollen or vaccine; secondly, environmental damage, or harm to natural resources; and finally, the possible negative impact of GM crops on other farmers. Each presents particular challenges of both prediction and appropriateness.

The first is the simplest. Human health might be injured in a number of ways, including by the consumption of GM food or by exposure to GM pollen. The meaning of damage will usually not be problematic. Debate about the impact of, for example, minor nutritional differences between crops would be as much evidential as conceptual, although the practical barriers should not be underestimated. A further significant barrier to successful claims may well be difficulties of proving causation. Many jurisdictions enjoy a mountain of very complex case law on this subject, but the basic position is simple: it is for the claimant to prove that the defendant's negligence or activity (depending on whether there is a strict liability or fault-based scheme) caused or contributed to his or her disease. It will often be impossible to prove that relatively common harms, such as heart disease or many forms of cancer, were caused by a defendant rather than background levels of risk. On the other hand, if a novel disease is overwhelmingly associated with a particular GMO, causation in this sense will probably speak for itself. The remaining causal difficulty will be identifying the appropriate defendant, if, for example, different firms produce similar or identical products.

Liability for environmental damage, as mentioned above, is specifically referred to as requiring legislative attention in the Deliberate Release Directive. Some of the possible harms are discussed in Chapter 2: reduced farmland biodiversity due to 'clean fields'; increase of land take and monocultures; herbicide or pesticide tolerance, with ecological as well as economic impacts; impacts of pest-resistant GMOs on non-target species; 'superweeds' that outcompete native wild plants, reducing biodiversity; and of course the spectre of unpredicted and unpredictable impacts in complex eco-systems. Generally, and notwithstanding certain rearrangements in different systems in recent years, tort law concentrates on interests in bodily integrity and property, and environmental quality is at most an incidental issue. The English common law, for example, ignores unowned environmental resources, and its focus on property and person can also lead to conceptual problems as to when contamination, or harm to plant or animal life, can be classified as 'damage' on which to base a claim. And many of the harms associated with GMOs are likely to occur on the GM farmer's own property – tort has nothing to say here.

The Environmental Liability Directive, however, is specifically concerned with 'environmental damage', including an imminent threat of environmental damage. Operators are required to take 'the necessary remedial measures' or 'the necessary preventive measures'.[92] There are three categories of environmental damage in the Directive: damage to protected species and natural habitats, water damage and land damage, each of which is further defined. The first is most obviously relevant for GMOs, and, although the others may apply, it makes sense to concentrate on this as an example. Protected species and habitats are those listed under EU nature conservation law,[93] plus, at the individual Member State's discretion, those protected under national law. Damage means 'a measurable adverse change in a natural resource or measurable impairment of a natural resource service which may occur directly or indirectly';[94] damage to protected species or natural habitats means 'any damage that has significant adverse effects on reaching or maintaining the favourable conservation status of such habitats or species'.[95] It is easy to see how some of the harms that may result from GM farming could be caught. The definition of 'environmental damage' in the Directive does not, however, fully address the possible environmental impacts of widespread cultivation of GM crops. As well as the detailed interpretation of 'significant adverse effects', the impact of GMOs is not limited to 'special' nature subject to legislative protection. 'Clean' fields, with no weeds and no pests, may have an obviously negative effect on farmland biodiversity,[96] but, unless the farm happens to be within a designated nature conservation area, farmland biodiversity is ignored. Nor would pesticide or herbicide resistance be readily encapsulated by the notion

---

92 Directive 2004/35, above n. 86, Articles 5 and 6.

93 Directive 1979/409 and Directive 1992/43, above n. 78. Negative impacts that have been authorised under this legislation do not attract liability.

94 Article 2.

95 In respect of habitats 'conservation status' is 'the sum of the influences acting on a natural habitat and its typical species that may affect its long-term natural distribution, structure and functions as well as the long-term survival of its typical species within, as the case may be, the European territory of the Member States to which the Treaty applies or the territory of a Member State or the natural range of that habitat'. Conservation status is 'favourable' when the following three conditions are met:

- its natural range and areas it covers within that range are stable or increasing,
- the specific structure and functions which are necessary for its long-term maintenance exist and are likely to continue to exist for the foreseeable future, and
- the conservation status of its typical species is favourable.

Directive 2004/35, above n. 86, Article 2(4)(a). A similar approach is taken in respect of species, Article 2(4)(b).

96 Hence the 'farm-scale evaluations' in the UK, discussed in Chapter 2.

of damage in the Environmental Liability Directive. This is no more easily embraced by normal tort liability. It will be recalled from Chapter 2 that a commercially prevalent modification is the insertion of a gene from the *Bt* bacterium, toxic to certain pests, into a crop. The current level of pest susceptibility to *Bt* toxin is a general public good. In a tort context, any resistance that develops is most likely to be conceptualised by reference to the impact on a particular farmer, which does not capture this broader public harm. Compensation will be assessed according to what is lost by the individual farmer rather than overall social costs, and used at the discretion of the farmer rather than directed to environmental benefits.

And that is if impacts on farmers are conceptualised as actionable 'damage' at all. Harm to farmers could take various forms, including loss of any market advantage that attaches to being 'GM-free', as well as the management costs of avoiding and removing GM material from land or coping with pesticide or herbicide resistance. But, in the absence of specific legislative intervention on GMOs, these harms have to be squeezed into the traditional framework of tortious damage to form the basis of a claim. The idea of physical damage to property is problematic. *Hoffman v Monsanto* in Canada involved an application by organic farmers for certification as a class under the Saskatchewan Class Action Act 2001, and one of the questions considered was the notion of damage in this context.[97] The farmers sought compensation because the spread of GM varieties means that it is now impossible to grow non-GM canola in Canada, and also because even those growing other crops incur costs removing GM canola volunteers. Specifically on damage, the loss suffered was characterised not as physical harm but as pure economic loss, in respect of which many jurisdictions (including Canada) impose limitations on recovery.[98] If the Courts refuse to characterise the harm as physical damage to property, a better option might be to argue that the defendants have unreasonably interfered with the use and enjoyment of the claimants' property, in the common law through the tort of private nuisance.[99] However, this is a notoriously unpredictable tort, and if the prospects of success are not negligible, nor

97 *Larry Hoffman, LB Hoffman Farms Inc and Dale Beaudoin v Monsanto Canada and Bayer Cropscience Inc* 283 DLR (4th) 190 (Saskatchewan Court of Appeal), [2005] SKQB225 (first instance). The plaintiffs have sought leave to appeal to the Supreme Court of Canada. The plaintiffs failed to meet the Saskatchewan requirements for a 'class' action. In respect of the examination of the causes of action by the first instance judge, the Court of Appeal saw 'no material error' in her approach, and indeed 'largely [agreed] with her analysis', para. 58.

98 *Hoffman*, above n. 97, para. 72, *per* Smith J.

99 See the sources cited above n. 89. *Hoffman*, above n. 98, did not consider this point in detail because the action was brought against the biotechnology industry rather than farmers; see below p. 136.

are they certain. When assessing harms to the use and enjoyment of property engendered by conflicting uses of land in private nuisance, the courts are necessarily involved in an evaluative task, driven by judicial understandings of reasonable or normal farming activities. Whether in respect of physical damage or interference with property, the existence of regulatory thresholds for 'adventitious' presence could be helpful in identifying damage, simply because they help to settle the point at which change becomes legally meaningful. This argument is particularly strong if regulatory thresholds are breached. It should also merit examination if the external regulator is a private organic certification body with a zero tolerance policy, but given the range of standards and the element of choice, this may be more difficult.

Many Member States are looking at legislation for liability in this area, which implies a legislative definition of actionable damage. The UK Government's current proposed statutory scheme of redress for non-GM farmers negatively affected by GM presence in their crops is tied to a 0.9 per cent threshold, and this is consistent with the Commission's approach to the subject.[100] The UK Government explicitly rejects special provision for losses brought about by 'voluntary standards or market-led decisions', including the removal of accreditation by an organic certifying body. It also rejects statutory compensation for reduced prices because a non-GM crop has been grown in the 'general locality of a GM crop, even though GM presence is below the required threshold', or for a 'precautionary decision' by a farmer not to grow a particular crop because of the possibility that it would be unacceptable due to proximity to GM crops.[101] Farmers in these situations will need to turn back to the underlying law of tort, taking us back to the difficulty of characterising the harm as 'damage' that counts.

The Parties

The identity of the appropriate claimant and defendant is by no means obvious in every case. The Environmental Liability Directive implies liability towards the state, and in fact the primary obligation is to restore, not to pay.[102]

---

100 See the Detailed Opinions under Directive 1998/34, above n. 30. DEFRA, above n. 52, para. 138 (all of these issues are open to consultation). Note that the German Genetic Engineering Act (as amended in 2006) does provide for liability below the 0.9 per cent threshold, in order to protect organic farming.

101 DEFRA, above n. 52, para. 148.

102 In the White Paper that preceded the Directive, public interest groups could take action if 'the State does not act at all or does not act properly', above n. 87, p. 22. Now, they are limited to judicial review of the regulator, Directive 2004/35, above n. 86, Articles 12 and 13.

The defendant is the 'operator', defined in the Directive as the person who 'operates or controls the occupational activity', 'occupational activity' being 'any activity carried out in the course of an economic activity, a business or an undertaking, irrespectively of its private or public, profit or non-profit character'.[103] Strict liability for all three types of environmental damage applies to 'operators' of a range of activities listed in an Annex to the Directive, including the deliberate release, transport or placing on the market of GMOs.[104]

The appropriate defendant in the case of environmental damage brought about by GM crops may well be the farmer, vet or transport company dealing with the GMO. In the case of widespread GM farming, this may mean that we are dealing with diffuse damage, which has always been difficult to regulate, and will not necessarily fall within the terms of the Directive: 'Liability is therefore not a suitable instrument for dealing with pollution of a widespread, diffuse character, where it is impossible to link the negative environmental effects with acts or failure to act of certain individual actors'.[105] The operative articles of the Directive provide that the Directive applies to 'pollution of a diffuse character' only when it is 'possible to establish a causal link between the damage and the activities of individual operators'.[106] That is of course a normal liability requirement. But a problem such as a diminution of biodiversity due to 'superweeds' looks far more like 'pollution of a widespread, diffuse character' if the defendant is a farmer than it does if the harm is the responsibility of the biotechnology industry. Imposing liability on one or even a number of farmers is not impossible, but if the use of GMOs becomes more normalised, pinning liability on individual farmers becomes more difficult.

The definition of 'operator' above could arguably be interpreted as applying to the company that sought authorisation of the GMO and then initially placed it on the market. Member States have the option of defining the operator as the person to whom: 'decisive economic power over the technical functioning of such an activity has been delegated, including the holder of a permit or authorisation for such an activity or the person registering or notifying such an activity'.[107]

This is not expressly designed to address the biotechnology industry, but reaching out to the biotechnology industry as defendants has its attractions. It converts superweeds, for example, into a normal, non-diffuse form of harm.

---

103 Directive 2004/35, above n. 86, Article 2(6) and (7).

104 Directive 2004/35, above n. 86, Article 3(1). In addition to strict liability for listed operations, 'occupational activities' that are not listed in the Directive are subject to fault-based liability in respect of damage to protected species or habitats.

105 Directive 2004/35, above n. 86, Recital 13.

106 Directive 2004/35, above n. 86, Article 4(5).

107 Directive 2004/35, above n. 86, Article 2(6).

And it is a commonplace of increasingly globalised agriculture that decisions of large corporations can have a profound impact on communities and environments very many miles away from where those decisions are made. In that context, responsibility to the local communities if things go wrong is hardly improper. It is also arguable that the industry is best placed to control for harms, and their liability would provide a strong incentive to do so. There is a choice to be made here.

In the case of personal injury or harm to farmers, the claimant is quite straightforwardly the individual suffering the harm. In either case, the claimant recovers for his or her own harm, not the more generalised environmental or public harms, and will use the funds recovered in his or her own interest. But the identity of the defendants is still not straightforward, particularly in the case of harm to farmers. The UK Government has consulted on three possibilities: GM farmers who have not complied with coexistence measures (which would of course introduce a fault-based system), all farmers growing GM crops, and GM seed companies.[108] If we could find some physical property damage or personal injury, a products liability claim would be the conventional approach, and in the EU context we would seek to bring an action against the manufacturer, including the biotechnology company that marketed the seed, although potentially embracing other defendants.[109] The main limitation on products liability is that the product in question must be 'defective', a notion that draws a line between, for example, a sharp edge on a knife, which is not a defect, and a sharp edge on a child's toy, which generally would constitute a defect.[110] There are certainly scenarios in which we could imagine a 'defective' GMO, especially if we have personal injury. This legislation is, however, not really designed to deal with the spreading of an organism in a way that is entirely expected of the product acting normally.

---

[108] DEFRA, above n. 52. Denmark notified its fund for the compensation of economic losses brought about by mixture with GM material under Article 88 EC on state aids. The Commission allowed this state aid, in part because it was financed by GM farmers. It was also found to be compatible with Commission policy on coexistence. See State Aid Case N568/2004, discussed in SEC (2006) 313, above n. 13.

[109] Directive 1985/374, above n. 91, Articles 1 and 3: 'Producer' means the manufacturer of a finished product, the producer of any raw material or the manufacturer of a component part and any person who, by putting his name, trade mark or other distinguishing feature on the product presents himself as its producer.

[110] Directive 1985/374, above n. 91, Article 6 provides: 'A product is defective when it does not provide the safety which a person is entitled to expect, taking all circumstances into account, including:

(a) the presentation of the product;
(b) the use to which it could reasonably be expected that the product would be put;
(c) the time when the product was put into circulation.'

Beyond statutory liability, it might be difficult in normal tort law to 'get at' the industry rather than individual farmers, as is demonstrated by the *Hoffman* litigation. Saskatchewan 'right to farm' legislation provided statutory protection from a nuisance action for farmers acting in accordance with 'normally accepted agricultural practices'[111] (raising all sorts of questions about normality) and hence the defendants in this case were two biotechnology companies providing GM canola seeds. As discussed above, the damage was not categorised as physical harm, and the defendants were held not to owe a duty of care to the plaintiffs in negligence. In addition, it was found that to hold a producer liable in nuisance simply because the *use* of a product causes a nuisance would be going too far. Everything from a radio to a radio mast is a potential nuisance in the wrong place or the wrong hands, and 'the implications of holding a manufacturer, or even inventor, liable in *nuisance* for damage caused by the use of its product or invention by another would be very sweeping indeed'.[112] Harm in private nuisance is *usually* a consequence of the use of land.

The Nature of Liability

The basic (somewhat overstated) division on the nature of liability is between strict and fault liability. Many of the significant issues in respect of the nature of liability are similar under these apparently very different rules. In particular, the role of 'foreseeability' of harm and the relationship between liability and prior regulation come up under a fault or a strict liability system, and these are the two questions that will be focused on here. Tort (like risk regulation) deals most readily with clearly calculable risks, whilst of course harm can be the result of issues along the whole scale of risk and uncertainty, including plausible but unproven risks and even mere suspicions of danger.

The Environmental Liability Directive provides for a core strict liability regime, but subject to a range of defences and exceptions,[113] including a 'state of the art' exception. It is possible to exclude from liability, at the discretion of the Member State, and provided that the operator is not 'negligent',

> an emission or activity or any manner of using a product in the course of an activity which the operator demonstrates was not considered likely to cause environmental damage according to the state of scientific and technical knowledge at the time when the emission was released or the activity took place.[114]

---

111 Agricultural Operations Act 1995.

112 See *Hoffman*, above n. 97, para. 122, *per* Smith J.

113 For example, damage caused by an act of armed conflict, Directive 2004/35, above n. 87, Article 4.

114 Directive 2004/35, above n. 86, Article 8(4)(b).

Strict liability thereby becomes subject to a form of foreseeability. This is in similar terms to a defence provided under the Products Liability Directive,[115] which was strictly interpreted by the ECJ in *Commission v UK*.[116] In the course of his opinion in this case, Advocate General Tesauro addressed the proper scope of the 'state of the art' defence:

> it is quite possible . . . that at the time when a given product is marketed, there will be isolated opinions to the effect that it is defective, whilst most academics do not take that view. The problem at this juncture is to determine whether in such a situation, that is to say, where there is a risk that is not certain and will be agreed to exist by all only *ex post*, the producer may still rely on the defence.[117]

He answered this question in the negative:

> the state of scientific knowledge cannot be identified with the views expressed by the majority of learned opinion, but with the most advanced level of research which has been carried out at a given time.[118]

The ECJ did not address this point in any detail, but its judgment did refer to 'scientific and technical knowledge, including the most advanced level of such knowledge',[119] reflecting the Advocate General's opinion. The ECJ is inclined then to take a narrow approach to this defence to strict liability. Producers of GM food and growers of GM crops would be wise not to ignore potential liability for risks that are predicted only by a minority of scientists. It is possible that the mavericks of today will turn out tomorrow to have been in possession of the most 'advanced knowledge'. We must however exercise some caution with *Commission v UK*. First of all, it was not a concrete liability claim, but an abstract consideration of the meaning of the Directive in a challenge to the adequacy of the UK's implementation. How the ECJ, or national courts, would respond to a real liability claim, with real minority science, really denied and perhaps convincingly by orthodox and mainstream science,

---

115 Which provides a defence if 'the state of scientific and technical knowledge at the time when [the defendant] put the product into circulation was not such as to enable the existence of the defect to be discovered', Directive 1985/374, above n. 91, Article 7(e).

116 Case C-300/95 *Commission v United Kingdom* [1997] ECR I-2649. The Commission unsuccessfully challenged the UK's implementation of the defence. The ECJ rejected the Commission's argument on the narrow basis that there was no evidence that the British courts would not interpret the defence in line with the Directive.

117 *Commission v UK*, above n. 116, para. 21.

118 *Commission v UK*, above n. 116, para. 21.

119 *Commission v UK*, above n. 116, para. 26.

is an open question. Neither the Court nor the Advocate General assesses the meaning of 'knowledge'. The most 'advanced' knowledge seems to be whatever turns out to be correct. This avoids any question of the credibility or cogency of the evidence at the time. In the regulatory context, as discussed in Chapters 2 and 3, famously, a preventive measure cannot be based on a 'purely hypothetical' approach to the risk, founded on 'mere conjecture which has not been scientifically verified'. The time for the application of the precautionary principle is when the risk, 'although the reality and extent thereof have not been "fully" demonstrated by conclusive scientific evidence, appears nevertheless to be adequately backed up by the scientific data available at the time'.[120] A certain level of demonstrable scientific evidence is demanded from regulators *before* a regulator can respond to risk. It might be strange in these circumstances if the courts required, on pain of a retrospective liability judgment, the producer to respond to hypothetical risks. Nor does *Commission v UK* consider how specific the state of the knowledge needs to be. Whether a broad prediction that a GMO may provoke an allergy suffices to take the particular allergic reaction that arises in the litigation out of the reach of the defence is not clear. We can readily anticipate some harm from GMOs, but not necessarily precisely the type of harm that ultimately occurs.

The boundaries of the state of the art defence are still uncertain, as is its application to GMOs. We need to think about who *should* bear the burden of ignorance or uncertainty. The uncertain and the unknown are the very things that raise concern in respect of GM agriculture, and indeed the very things that 'regulation' struggles to address. Strict liability with no foreseeability or state of the art defence is a way of ensuring that these unknowns are the responsibility of the industry. The significant costs of uncertainty are otherwise imposed by the new technology on existing enterprises, consumers or the environment. This seems quite straightforward, but is far from easy. The 'state of the art' defence is designed to protect innovators on the assumption that innovation is in the public interest. It is because there is no consensus on the public interest in the innovations provided by the biotechnology industry that the application of the defence is so contentious and particularly far-reaching. Underlying an apparently small technical legal choice are completely different ways of understanding the technology. This is a deeply political question that ought to have been decided as part of the overall regulatory framework.

We are dealing here with a heavily regulated area, and it is likely that the GMOs in question have been authorised under the EU's risk regulation regime, and may also be used in accordance with a coexistence regime set out

120 Case T-13/99 *Pfizer Animal Health SA v Council* [2002] ECR II-3305.

at national level. An important question is the relationship between liability and regulation, particularly the extent to which prior authorisation and risk assessment might provide some form of defence to liability (or, in a fault-based scheme, the circumstances in which a 'reasonable person' might fail to comply with regulation). Harm resulting from 'an emission or event expressly authorised by, and fully in accordance with the conditions of, an authorisation conferred by or given under applicable national laws and regulations' can be excluded from liability under the Environmental Liability Directive, at the discretion of the Member State, and provided that the operator is not negligent.[121]

The appropriate relationship between a regulatory scheme and liability is far from obvious. We might for the sake of argument assume that the granting of authorisation implies a consensus that the GMO is in the public interest, but that does not necessarily say anything about where the costs of things going wrong should lie. Certain risks are inherent in even the heavily regulated development of GM farming – the distribution of the benefits and burdens of those risks is important. Further, a compliance defence places a great deal of faith in the ability of the regulator to get things right first time, a faith not often apparent when regulation is assessed in its own terms. And we are not simply concerned with 'getting it right' in the sense of predicting harms, but also with addressing the acceptability of those harms. In the current context, that includes the question of whether the harms addressed by regulation are the right types of harm. Again, this is a difficult, and deeply political, decision. And again, because the particular innovation is so contentious, so is the detail of liability. The questions include whether the regulatory authority has already made a determination of the public interest, whether that prevents the courts revisiting questions of public interest, and whether that determination of the public interest trumps any private interest of the claimant. The Food and Feed Regulation provides explicitly that 'the granting of authorisation shall not lessen the general civil and criminal liability of any food operator in respect of the food concerned',[122] clearly an intention that authorisation shall not preclude liability in respect of food, although, if 'general' liability rules are reluctant to impose liability for authorised activities, it may not take us very far.

The Commission, without apparently giving the subject an enormous amount of consideration, appears to think that strict liability would be inappropriate in respect of economic harms done to farmers. Many jurisdictions justify their use of strict liability by reference to 'hazardous' activities, or some

121 Directive 2004/35, above n. 86, Article 8(4)(a).
122 Regulation 1829/2003, above n. 12, Article 7(7).

variation on that theme.[123] The Commission thinks that it 'would not be appropriate' to classify GM farming as a 'hazardous activity', specifically because authorisation means that a GMO is 'to be considered as safe'.[124] Aside from being quite extraordinarily complacent about the complexities of regulating GMOs,[125] it must be obvious that most activities that we might classify as hazardous and subject to strict liability (nuclear power stations or oil pipelines for example) are also subject to regulation to ensure as far as possible their safety. The Commission should either give considerably more thought to this sort of liability question or intrude far less into national decisions. Similarly, the Commission seems elsewhere to assume that a farmer complying with 'good farming practice' could not conceivably be made liable for resulting harm.[126] In this, the Commission simply ignores the most difficult parts of the questions it purports to answer, that is, the role of strict liability, the relationship between regulation and liability, and the distribution of risks and costs through different legal mechanisms.

We should also note that in the absence of an express decision, even beyond any question of fault or defences, the fact of government approval could inhibit judicial intervention in agriculture. The Saskatchewan Court of Appeal in *Hoffman* agreed with the first instance court that government approval provided 'sound policy reasons' for denying the existence of a duty of care in respect of contamination.[127] This is a specifically common law point, but the judicial philosophy apparent in this hands-off approach may well apply more generally.

Conclusions on Liability

The liability situation for any unwanted effects of agricultural biotechnology is far from certain. The openness of the English common law, for example,

---

[123] This is to some degree the case in the English common law, with the approach to 'natural use' in *Rylands v Fletcher* liability, and specific statutes apply strict liability to, for example, nuclear installations or marine oil pollution, following international law.

[124] European Commission, Detailed Opinion under Directive 1998/34/EC Laying Down a Procedure for the Provision of Information in the Field of Technical Standards and Regulations [1998] OJ L 204/37, above n. 58, para. II(6).

[125] Decisions under Directive 1998/34 are taken by DG Enterprise and Industry.

[126] See European Commission, Detailed Opinion under Directive 1998/34/EC Laying Down a Procedure for the Provision of Information in the Field of Technical Standards and Regulations [1998] OJ L 204/37, *Notification 2004/133/D (Draft Act Reorganising legislation concerning genetic legislation)*. The Commission seems to be concerned with causation as much as fault, but it is very unclear.

[127] *Hoffman*, above n. 97, para. 59.

leaves the allocation of liability in at least certain respects largely to the discretion of judges. Without doing any violence to the authorities, protection might be given to organic *or* to GM farming. Whilst one can expect the courts to seek to apply legal principles in an even handed manner, unseen assumptions and values are very likely to influence a judgment as to whether, for example, GM farming is a 'natural' or 'reasonable' use of land, whether organic farming is a 'sensitive' use of land, whether a GM farmer is behaving 'reasonably' in his or her coexistence measures.[128] The ways in which liability distributes costs and risks is deeply political and the detail of any final scheme is crucial. The right place to have resolved these issues would have been in the negotiation of the regulatory framework, forming part of the social arrangements that determine the acceptability (or otherwise) of agricultural biotechnology. It is because consensus on the benefits of technological 'progress' has fallen apart in the case of GMOs that arranging liability is so difficult. Moreover, as discussed in Chapter 5, the biotechnology industry seeks very high levels of control over its products when that is in its economic interests. The industry supported in that by the courts, and the mismatch between that control and its resistance to legal responsibility if things go wrong is rather striking.

LABELLING

One of the objectives of the Food and Feed Regulation is to provide a 'high level of protection of . . . consumer interests'.[129] The primary tool for the protection of consumer interests is mandatory labelling of GMOs, which 'enables the consumer to make an informed choice and facilitates fairness of transactions between seller and purchaser'.[130] This is a crucial component of GM regulation, enabling choice on the basis of a criterion that would not otherwise be visible to purchasers. Whilst the ability to participate in the market is progress – the bulk of the biotechnology industry's direct customers are farmers, and members of the public have only belatedly been seen as consumers of GMOs – it is nevertheless a limited response to a very complex question. Not only is the real 'choice' available to the consumer limited by gaps in the labelling obligations, but the potential for individual consumption decisions to respond fully to the collective, political controversy around issues such as GMOs is somewhat overplayed.

---

[128] See generally the citations at n. 89 above.

[129] Regulation 1829/2003, above n. 12, Article 1. Note, though, that Article 153 (on consumer protection) is not one of the bases of the Regulation.

[130] Regulation 1829/2003, above n. 12, Recital 17.

As discussed in Chapter 3, GMOs have been subjected to safety assessment along the lines of risk to the environment and to human health, and may have also been assessed against 'other legitimate factors' in the case of food or feed. The fact of authorisation does not however mean that consumers will accept GMOs. The basic adequacy of the risk assessment may be disputed, particularly given the prominence of uncertainty and ignorance in this area, and some consumers may decide that the risk of the unknown is not worth taking. Some consumers may wish to reject GMOs on other political or ethical grounds or may be responding to those developing the technology as much as to the technology itself. Misgivings about the distribution of the costs and benefits of GMOs, a wish to resist corporate control of agriculture, and a biocentric or religious perspective that doubts the wisdom of the interference with nature implied by agricultural biotechnology could all affect consumption decisions.[131] Mandatory labelling of GMOs allows consumers to reject GMOs for any reason, or none. If much of this book is about finding space for the expression of the full breadth of issues around GMOs, labelling means that we have some hope in the market.

The industry has always resisted calls for mandatory labelling, arguing that there are no legitimate grounds for distinguishing between GM and non-GM products.[132] This implies that the 'irrational' exercise of choice, going beyond proven harm to health or the environment, should be left solely to voluntary labelling.[133] The argument that government support for consumer choice should, like regulation, be predicated on tangible harms indicates the intensity of resistance to the broad nature of concerns about GMOs. Such a massive restriction of the opportunities to explore concern about GMOs would increase the pressure for those concerns to be brought forward into the explicitly regulatory process, including probably coexistence, which would have to justify independently its use of the handy 0.9 per cent threshold. We should also note that the ECJ consistently emphasises the role of labelling as a tool of information, not as a tool of environmental protection.[134]

---

[131] See the more detailed discussion of the concerns surrounding GMOs in Chapter 2.

[132] Joyce Tait and Joanna Chataway, 'The Governance of Corporations, Technological Change, and Risk: Examining Industrial Perspectives on the Development of Genetically Modified Crops' (2007) 25 *Environment and Planning C: Government and Policy* 21, p. 32. This has been raised in the WTO context as well; see Chapter 6.

[133] See the discussion in Christopher Hilson, 'Information Disclosure and the Regulation of Traded Product Risks' (2005) 17 *Journal of Environmental Law* 305.

[134] For example *Codacons*, above n. 24; Case C-316/01 *Glawischnig v Bundesminister für soziale Sicherheit und Generationen* [2003] ECR I-5995 (denying the applicability of Directive 2003/04 on Public Access to Environmental Information [2003] OJ L 41/26 to GMOs).

More pragmatically, industry has always denied, right back to the efforts of some supermarkets in the 1990s to avoid GMOs, that it is possible to differentiate in a reasonable way between (especially) commodity crops. Similarly, industry argues that segregation raises the costs of both GM and non-GM products, whilst voluntary GM-free labels place the cost on those seeking GM-free produce. The preference for voluntary as opposed to mandatory labelling is, like so many of these supposedly technical debates on the detail of GMO regulation, a reflection of fundamental and inconsistent understandings and expectations of the genetic nature of products. It depends on whether we think that the consumer who prefers non-GM food is seeking special characteristics, or the norm. More practically, seeking to pursue GM-free status without regulatory support places an enormous, possibly unmanageable, burden on the GM-free sector. Thinking about distribution forces us to ask how much of that burden should be carried by the sector introducing novel forms of production.

The legislation assumes that it will be unproblematic to trace, and implicitly also to separate, products. It imposes obligations to ensure 'the ability to trace GMOs and products produced from GMOs at all stages of their placing on the market through the production and distribution chains'.[135] Traceability is achieved by information. The information that products 'contain or consist of', or in the cases of food and feed have been 'produced from', GMOs, as well as the 'unique identifier' of the relevant 'transfer event' has to be transmitted and held at each stage of marketing.[136] There should be a continuous paper trail for every regulated product of biotechnology as it passes through the EU. Importers and third-country governments continue to object that traceability puts an undue burden on GM food. By contrast, the Commission attributes low levels of GM food on the EU market to market demand rather than burdensome regulation.[137]

Labelling is mandatory for all products covered by the Food and Feed Regulation ('GMOs for food use', 'food containing or consisting of GMOs' and 'food produced from or containing ingredients produced from GMOs'), and for all other 'products consisting of or containing GMOs'.[138] The labelling obligations are very clearly concerned with the characteristics of production,

---

135 Regulation 1830/2003, above n. 51, Article 3(3). The objectives of the Traceability and Labelling Regulation include 'facilitating accurate labelling', alongside safety considerations, 'monitoring the effects on the environment, and, where appropriate, on health, and the implementation of the appropriate risk management measures including, if necessary, withdrawal of products', Article 1.

136 Regulation 1830/2003, above n. 51, Article 4.

137 European Commission, *Report on the Implementation of Regulation (EC) No. 1830/2003 concerning the Traceability and Labelling of Genetically Modified Organisms and Amending Directive 2001/18/EC* COM (2006) 197 final, pp. 4–5.

138 Regulation 1830/2003, above n. 51.

not just the characteristics of the product itself. This is especially true of food and feed: the labelling obligations extend even to heavily processed foods ('produced from' GMOs), and so are not dependent on the presence of GM DNA or protein in the final product.[139] As Douglas Kysar puts it, consumers have legitimate 'preferences for processes' whether or not the process by which a product is made is reflected in the characteristics of the product. To prevent the exercise of those preferences would be especially strange given contemporary privileging of the role of markets in public policy.[140] It should be apparent that this approach demands considerable vigilance from regulators, but it is important that the legislation has acknowledged that the *process* of agricultural biotechnology (and not just the product) is of interest to the consumer, and renders this visible.

Mandatory labelling of GMOs is a progressive and important element of their regulation in the EU. Whilst the labelling obligations are in many respects broad, and certainly controversial in their breadth, however, there are gaps in the legislation. First, the concern with production methods does not extend to non-food or -feed GMOs. Under the Traceability and Labelling Regulation (applying labelling obligations to non-food or -feed GMOs[141]) there is no obligation to label products *produced from* a GMO, for example cotton clothes. Secondly, whilst the Food and Feed Regulation does include products produced *from* a GMO, it does not apply to food and feed produced *with* a GMO, for example, the meat or milk of a cow fed on GM feed.[142] These two exclusions sit uneasily with the decision to include food 'produced from' GMOs, the justification of which refers to 'the demands expressed in numerous surveys by a large majority of consumers', stating that such labelling 'facilitates informed choice and precludes potential misleading of consumers as regards methods of manufacture or production'.[143] This is yet another rather awkward compromise between a desire to facilitate marketing of GMOs and a desire to control it. Given the potential, and actual,[144] size of the market for

---

139 Scientific testing cannot identify whether there once was a GMO used in the production of this food and feed – the only thing identifying this product as a subject of regulation is the paper trail.

140 Douglas A. Kysar, 'Preferences for Processes: The Process/Product Distinction and the Regulation of Consumer Choice' (2004) 118 *Harvard Law Review* 525.

141 Regulation 1830/2003, above n. 51.

142 Regulation 1829/2003, above n. 12, Recital 16. The feed itself has of course been authorised.

143 Regulation 1829/2003, above n. 12, Recital 21.

144 The vast majority of the GM market in the EU is for animal feed, and GM cotton is also a huge worldwide market. The bar on use of GMOs in organic production extends to animal feed and processed non-food.

animal feed and for processed non-food GMOs, consumer influence over the cultivation of GM crops is significantly restricted. If GM farming is a change in the world that certain consumers wish not to support, they need to expend considerable effort to identify and avoid some of the products that do indeed support GM agriculture.

A third practical limitation on the obligation to label relates to the threshold of GM presence below which labelling is not required. As discussed above, labelling is not required if a food, or an individual ingredient in a food, contains up to 0.9 per cent GM material, provided that the presence of GM material is 'adventitious or technically unavoidable'.[145] Whilst this has serious implications for coexistence (discussed above), in the narrow context of the regulation of labelling it is clearly a limit on the ability of consumers to choose. The more extensive adventitious presence becomes, the greater the limitation. The introduction of thresholds constitutes a pragmatic recognition that, once GMOs are on the market, they will be virtually impossible to avoid entirely, and the low thresholds do require considerable efforts at segregation, the feasibility of which was initially denied by producers of grain. The consistency of thresholds with the consumer choice rhetoric is nevertheless somewhat tenuous. Pursuit of the GM-free option (or the GM-free 'consumer choice') is in a practical sense abandoned.

Advocate General Léger's Opinion in *Codacons*, a case in which an Italian court had referred a question to the ECJ as to whether labelling thresholds apply to baby food, gives us a clue as to the potential judicial approach to labelling.[146] He examined whether the absence of a label on baby food containing traces of GM material was misleading to consumers, applying the standard of 'the presumed expectations of an average consumer who is reasonably well informed and reasonably observant and circumspect'.[147] The Advocate General took the view that such a consumer would indeed 'expect' a small proportion of GMOs to be present in unlabelled baby food:

> The contamination of the environment by GMOs is a well-known phenomenon and is regularly reported in the media. . . . the contamination of the environment by GMOs is a fact which can hardly be unknown to the average consumer who is reasonably well informed and reasonably observant and circumspect. It may also be presumed that such a consumer may expect that foodstuffs for infants and young

---

145 Regulation 1829/2003, above n. 12, Article 12(4).

146 *Codacons*, above n. 24. The case was taken under the old legislation (Regulation 258/97 Concerning Novel Foods and Novel Food Ingredients [1997] OJ L 043/1), under which the threshold was 1 per cent.

147 *Codacons*, above n. 24, Advocate General's Opinion, para. 78.

> children will not be free of slight impurities or foreign substances, in spite of the efforts of the manufacturers to prevent the inclusion of material derived from such organisms in those products.[148]

Whilst the Advocate General recognised that 'in some cases, consumers may be unaware of this fact and may thus be misled by the absence of any reference to GMOs', he took the view that 'that risk remains minimal'.[149] The substance of the decision is perhaps not surprising, given the failure to provide any evidence that GMOs are harmful to infants and the absence of an explicit exception in the legislation for baby food. But the extreme approach to the 'average consumer' opens up the paradox of limiting choice by the very same means by which the choice is provided. Léger AG's prodigious 'average' consumer is perhaps most likely to take political action through consumption decisions on the basis of information. But this consumer is also highly aware of the restrictions on the choice on offer. Not only is this consumer perhaps discouraged from exercising choice, but necessarily is less needy of the protection of legislation – the role of labelling in equalising (just slightly) the informational imbalances between consumer and producer is ignored.

The intimately connected labelling, coexistence and consumer choice elements of the GMO regime supplement efforts to respond to the range of public concerns about GMOs through authorisation. Indeed, notwithstanding the very strong rhetorical commitment to 'public participation' in environmental and more general EU governance, the primary route for participation in the regulation of GMOs seems to be as a 'consumer', with only limited opportunities for public comment in the regulatory process.[150] Consumption is indeed capable of being a political action, or an opportunity for collaborative political engagement through private decisions.[151] Trying to draw clear lines between consumer preferences and citizen values is probably impossible,[152]

---

148 *Codacons*, above n. 24, Advocate General's Opinion, paras 81–2.

149 *Codacons*, above n. 24, Advocate General's Opinion, para. 83.

150 See Chapter 3.

151 See Michelle Micheletti, Andreas Follesdal and Dietlind Stolle (eds), *Politics, Products and Markets: Exploring Political Consumerism Past and Present* (Transaction Publishers, 2004); Martin Daunton and Matthew Hilton (eds), *The Politics of Consumption: Material Culture and Citizenship in Europe and America* (Berg, 2001).

152 Mark Sagoff, *The Economy of the Earth: Philosophy, Law and the Environment* (Cambridge University Press, 1988), provides classic and hugely important observations on the distinction between consumer preferences and citizen values, and argues against the over-privileging of the former. For criticism of Sagoff's line-drawing, see Carol M. Rose, 'Environmental Faust Succumbs to Temptations of Economic Mephistopheles, or Value by Any Other Name is Preference' (1989) 87 *Michigan Law Review* 631; on blurred lines between citizens and consumers, see for

and the politicisation of consumption implied by a demand for information should be taken seriously.[153] Nevertheless, as well as the arguably cramped meaning of 'choice' in this case, there are other more general difficulties with the rhetoric of choice.

If an individual is unhappy about GMOs, or fears their health impact, that individual should be able to avoid consuming GMOs. This, however, assumes the paradigm of the informed, active, empowered, well-informed and well- resourced consumer – others are potentially excluded. And, although consumers can have a collective impact, collective action through consumption is problematic. Consumers are generally individualised, with individual rights and an economic role. Consumer choice addresses an individual's concern for his or her own health or own moral probity. A consumer who wishes to be more precautionary than EU law allows regulators to be may be so; a consumer who wishes to take into account factors beyond those addressed by regulator, may do so. However, some of the mooted health impacts of GMOs are of concern species-wide, not just for the individuals who consume them. Similarly, an individual may be ethically concerned not simply about his or her own impact on the environment, but with the *overall* impact or risk of GMOs. In either case, the instrumental effect of individual action is limited. The cost of rejecting GM may cease to be worthwhile on an individual basis, whilst that same cost would be worthwhile if collectively undertaken through regulation. It is likely that people will base their decisions in part not only on their own views of the rights and wrongs of GMOs but also on what they think other people will do. It would be surprising if it were otherwise, because these are after all collective and not individual decisions.

This leads to the rather familiar prospect of an individual's short-term consumption decisions being used to justify positions that the same individual would not support politically. Essentially public and collective goods are forced into an individualised framework, fudging government responsibilities. We are not all equal before the market. Without disagreeing that consumption can sometimes be a political act, it is not always hypocritical to adapt to the frameworks set by the dominant forces in a society. The market power of the multinational biotechnology industry is rather impressive, and it is not difficult to envisage a policy of aggressive pricing and marketing overwhelming political preferences expressed individually, or at least creating consumer

---

example Lizabeth Cohen, 'Citizens and Consumers in the United States in the Century of Mass Consumption' in Daunton and Hilton, above n. 151.

153 Martin Daunton and Matthew Hilton, 'Material Politics: An Introduction' in Daunton and Hilton, above n. 151, especially pp. 14 and onwards.

resignation in the face of the apparently inevitable.[154] The purchase of GMOs does not imply the wholehearted approval of the technology by the person doing the purchasing.[155]

How any post-authorisation public mistrust of GMOs will play out in the marketplace is unpredictable, and will not be straightforwardly reflected in consumption decisions. Consumers make choices on the basis of a range of factors. Their choices are 'embedded in culture', implicating all kinds of social, political and moral value systems.[156] The assumption of much policy on GMOs seems to be that consumers can dictate the success or failure of agricultural biotechnology in the market place. This is, though, highly debatable. The fact that an individual has consented (chosen even) to consume GMOs does not signal an end to the argument, because the argument goes beyond those issues subject to choice in the market, and beyond effects on the individual consuming the good.[157] This is about the way we run our agricultural systems, not about competing brands of cornflakes.

CONCLUSIONS

The post-authorisation regulatory framework for agricultural biotechnology is still open. At the moment, however, the conceptual limitations on authorisation are being repeated afterwards. Just as an overwhelmingly scientific focus in risk assessment can be an attempt to silence other voices, so the highly technical legal focus after authorisation has obvious potential to diminish the richness of the debate. A more careful approach should be possible, allowing at least an examination of the genuine public interests caught up particularly in coexistence and liability.

---

[154] Daphna Lewinsohn-Zamir, 'Consumer Preferences, Citizen Preferences and the Provision of Public Goods' (1998) 108 *Yale Law Journal* 377, talks of 'hopelessness'.

[155] See Robin Grove-White, Philip MacNaghten, Sue Mayer and Brian Wynne, *Uncertain World: Genetically Modified Organisms, Food and Public Attitudes in Britain* (CSEC & Unilever, 1997). See also Nuffield Council on Bioethics, *Genetically Modified Crops: the Ethical and Social Issues* (Nuffield Council on Bioethics, 1999), Chapter 5, for more discussion.

[156] Christer Berglund and Simon Matti, 'Citizen and Consumer: the Dual Role of Individuals in Environmental Policy' (2006) 15 *Environmental Politics* 550, p. 551.

[157] There are resonances here with the relevance of consent in respect of medical biotechnology – the fact that the individual receiving controversial treatment is not the only concern if the procedure in question poses risks or concerns that extend beyond that individual.

The sharp separation of the legal environment in which GMOs will be grown and marketed from the questions raised during authorisation is difficult to maintain. The current framework for coexistence leaves some key issues entirely unaddressed, and responds poorly to others. The future of organic agriculture (as currently understood), in particular, looks bleak and it is not clear why a redefinition of what we mean by 'organic' and 'GM-free' is a positive conclusion to the debate on coexistence. The approach to be expected on liability is largely obscure to the public and in many respects even to experts, when it is central to questions of responsibility that are in turn fundamental to the acceptability of the regulatory framework. Mandatory labelling does, however, allow the market to become a space for the expression of individual and collective values in respect of GMOs. The consumer choice mantra, however, assumes that the choices available are meaningful, when the alternatives are in fact severely limited.

The entire policy discussion seems to assume that widespread GM agriculture is an objective and unavoidable fact to which other forms of agriculture must adapt, rather than a choice. With the gradual and almost constant changes in the regulation of GMOs over the past two decades, we have seen the development of an extremely sensitive and elaborate arrangement of central and national responsibilities. Precisely how that will play out is still to be explored, but the significant role of the European Food Safety Authority, together with the inability of Member States to reach agreement in committee or Council, has led to a centralisation of risk regulation, in the interests of markets. The Commission's approach to coexistence suggests a concern about the repatriation of discretion by the back door. The tight control of GMOs may turn out to be a poisoned chalice for the internal market, especially at a time of increased popular scepticism about the legitimacy and effectiveness of EU law.

5. Living with GMOs (2): ownership

INTRODUCTION

The 'regulation' of agricultural biotechnology is frequently presumed to begin and end with the procedure for the authorisation of a product or process, generally now through a risk-focused framework.[1] The dominance of this risk paradigm can, however, distract attention from other significant issues, and, more realistically, regulation stretches both forwards and backwards from authorisation. The central contention of this chapter is that the regulation of GMOs is intimately connected with the rights, interests and liabilities awarded and arbitrated through intellectual property law. Intellectual property rights contribute to the shaping of the relationship between the biotechnology industry and those it affects, including farmers, potential consumers, and broader publics, both locally and globally. As such, intellectual property demands the close attention of those interested in the regulation of agricultural biotechnology.

Whilst it tends not to be cast directly in a regulatory role by patent lawyers, it is conventional to see patent law discussed in terms of influencing behaviour,[2] and indeed the economy. In particular, the importance of patent rights in stimulating research and development of new technologies, in turn crucial for a prosperous economy, is a widespread and deep-seated belief in EU policy and law.[3] Holding patents is seen as a measure of success, for countries[4] and regions as well as for companies, and there is clearly a perception, at least on the part of the European Commission, that the EU has lagged behind its

1 Most obviously the legislation discussed in Chapter 3. Elizabeth Fisher, 'Risk and Environmental Law: A Beginner's Guide' in Benjamin Richardson and Stepan Woods (eds), *Environmental Law for Sustainability* (Hart Publishing, 2006), discusses the way in which environmental regulation in particular turned into a question of 'risk' in the last two decades of the 20th century.

2 See Brad Sherman, 'Regulating Access and Use of Genetic Resources: Intellectual Property Law and Biodiscovery' [2003] *European Intellectual Property Review* 300; Graham Dutfield, *Intellectual Property Rights and the Life Science Industries: A Twentieth Century History* (Ashgate, 2003), Chapter 2.

3 See for example European Commission, *Mid-term Review of the Strategy on Life Sciences and Biotechnology* COM (2007) 175 final, section 5.

4 See the discussion in Andrew Barry, *Political Machines: Governing a Technological Society* (The Athlone Press, 2001), Chapter 5.

competitors.[5] The EC's Biotechnology Directive provides that the products of biotechnology are patentable,[6] a clarification of the reach of patent rights that is presented as crucial to the economic development of Europe. The pervading premise of the patent system that innovation is a general public good and to be encouraged has, however, broken down in respect of agricultural biotechnology, especially on its current profit-driven trajectory. In a comment revealing of the interests served by the patenting system, the Commission asserts the need for 'Regular assessments . . . on whether the patent regime satisfies the needs of researchers and companies'.[7] Patent law is one aspect of the regulatory effort to tread a line between encouraging a potentially valuable technology and controlling that technology in the public interest. To examine only the control through the authorisation process would be to miss half of the picture.

This chapter does not purport to provide a detailed legal analysis of patent law and its application to agricultural biotechnology.[8] Instead, I begin with the briefest of outlines of the patent system, before examining particular controversial areas that point up the ways in which the patent system exercises a broader influence. Specifically, this chapter first considers the distributive impacts of patent rights. Social benefits developed by the ingenuity of farmers over millennia, or naturally occurring in areas of high biodiversity, are developed and commodified through patent rights. This is an issue with particular, although not exclusive, resonance at the global level, influencing the allocation of risks and benefits between developing and developed countries. As intellectual property standards cross national borders (just as technologies are increasingly 'globalised'), the stakes are raised, and the space for political

---

5 Going back as far as European Commission, *Biotechnology in the Community* COM (1983) final. See also the *Mid-term Review*'s comparison with, especially, US patenting, above n. 3 para. 2.1, and E. Richard Gold and Alain Gallochat, 'The European Biotech Directive: Past as Prologue' (2001) 7 *European Law Journal* 331.

6 Directive 1998/44/EC on the Legal Protection of Biotechnological Inventions [1998] OJ L 213/13.

7 European Commission, *Life Sciences and Biotechnology – A Strategy for Europe* COM (2002) 27 final, p. 23; see also *Mid-term Review*, above n. 3, 'Europe's dedicated biotech companies are mostly SMEs with limited resources whose growth and economic sustainability are held back by three main constraints: Europe's fragmented patent system, the insufficient supply of risk capital and shortcomings in the cooperation between science and business', para. 5.2.

8 See for example Lionel Bently and Brad Sherman, *Intellectual Property Law* (Oxford University Press, 2001), Chapter 14; Margaret Llewellyn and Michael Adcock, *European Plant Intellectual Property* (Hart Publishing, 2006); Philip Grubb, *Patents for Chemicals, Pharmaceuticals and Biotechnology: Fundamentals of Global Law, Practice and Strategy* (Oxford University Press, 2004); Tanya Aplin and Jennifer Davis, *Cases and Materials on Intellectual Property* (Oxford University Press, 2008, forthcoming).

controversy in this highly technical (legally and scientifically) area increases. Turning to the local level, and building on some of the themes developed in the previous chapter, I explore the impact of patent protection on the coexistence of different forms of agriculture. Patent law exercises a fundamental influence, with clear distributive impacts, over relationships between farmers, potentially even subverting the direct regulation of coexistence. And then, moving on from explicitly distributional questions, the debate about the appropriate scope of ownership of the living products of biotechnology is examined. The contest over the reach and content of patent rights is also a contest over the sort of economic and social environment in which we wish to live.[9] Patenting is primarily a tool for commerce, providing ownership in the results of research and rendering them more amenable to trade and exploitation.[10] Perhaps surprisingly, however, patenting provides a rare opportunity for the moral scrutiny of the products of agricultural biotechnology, through the explicit provision in the European system that inventions the exploitation of which would be contrary to 'ordre public' or morality are not patentable. The morality exception is the final topic of this chapter.

Each of these elements of intellectual property is highly significant in regulatory terms, at least for its influence over the relationships between different parties before and after authorisation and over the distribution of risks, costs and benefits. As such, intellectual property provides part of the context within which the more explicitly 'regulatory' elements of the framework apply to agricultural biotechnology.

PATENTS

Disagreement about the appropriate nature of ownership in agriculture is not new, and there was considerable debate about the privatisation of plant resources (in our case especially food resources) in early post-war Europe. At that time of scarcity, it was thought that plant breeders were performing tasks of urgent public interest, and that, although they lacked an 'invention' qualifying for a patent, some form of intellectual property protection was needed. In addition to physical ownership of a plant or seed, a distinct right to intellectual property prevents others from copying the physical entity, allowing reward for intellectual effort. The perceived need for protection was, however, matched by an awareness of the more general needs of agriculture and food

---

9 For discussion, although not in the context of biotech, see Barry, above n. 4, Chapter 5.

10 Grubb, above n. 8, Chapter 1, for example, makes clear the connection between patents and commerce.

production.[11] The protection of new plant varieties through 'plant variety rights',[12] granted over plant groupings that are 'distinct, uniform and stable', was designed in such a way as to balance support for innovation and support for traditional practices of food production. Margaret Llewellyn and Michael Adcock argue that plant rights were initially intended to be distinct from patents, specifically because of concerns about control of food crops, but that over time the balance of protection has shifted towards the developer as plant rights increasingly resemble patent rights.[13] This chapter concentrates on patents rather than plant variety rights, as patents are the most common route for protection of biotechnological inventions.

A patent provides a monopoly right to exploit an 'invention' for a specified period (in most cases, 20 years[14]) in the jurisdiction that grants the patent. This monopoly is generally presented as a reward for the research and development of an invention,[15] and also in exchange for disclosure of the invention. Without the monopoly rights granted by patent, it is argued, there would be no incentive for an inventive sector in the economy, as it would be more rational to free-ride on the research and development of others, leading to an overall reduction in research activity. One purpose of the EU's intervention in patenting of biotechnological inventions (the Biotechnology Directive[16]), highlighted among other purposes by the ECJ, is 'to promote research and

11 See Llewellyn and Adcock, above n. 8.

12 Regulation 2100/1994/EC on Plant Variety Rights [1994] OJ L 227/1 is based on the International Union for the Protection of New Varieties of Plants (UPOV). For a useful discussion of the difficulties of defining 'plant variety', as well as the legal approach, see Margaret Llewellyn, 'The Patentability of Biotechnological Material: Continuing Contradiction and Confusion' (2000) 22 *European Intellectual Property Review* 191; G01/98 *Novartis/Transgenic Plant* [2000] EPOR 303; Llewellyn and Adcock, above n. 8.

13 Llewellyn and Adcock, above n. 8. Note the dissent in the US Supreme Court decision in *JEM Ag Supply Inc v Pioneer Hi-Bred International Inc* 534 US 124, 122 S Ct 593, arguing that to allow patent protection for plants destroys the exemptions (relating to seed saving and research) to protection for plant varieties, p. 611.

14 'Plant protection products' can be granted an extension to reflect the time taken to get regulatory approval; see Regulation 1610/1996/EC on the Creation of a Supplementary Protection Certificate for Plant Protection Products [1996] OJ L 198/30. Genetic modifications that serve the same purpose (e.g. to act as an alternative to a pesticide or artificial fertiliser) are also subject to protracted authorisation processes; we might expect extensions to be sought.

15 For detailed discussion of the justifications of patent law, and the longstanding nature of the debate, see Fritz Machlup and Edith Penrose, 'The Patent Controversy in the Nineteenth Century' (1950) 10 *The Journal of Economic History* 1; also the sources cited above n. 8.

16 Directive 1998/44, above n. 6.

development in the field of genetic engineering in the European Community'.[17]

The applicant for a patent must prove that the invention is novel, includes an inventive step, and is capable of industrial application and of adequate disclosure by description. These basic criteria apply through much of the world, although different approaches may mean that an application for a patent in the same 'invention' fares differently in different jurisdictions.[18] Each of the criteria for granting a patent has been contentious in its application to the products of agricultural biotechnology,[19] and these are terms of law rather than science or everyday language.[20] An invention is to be distinguished, sometimes with difficulty, from a 'discovery', which is not patentable. 'Novelty' requires that the invention 'does not form part of the state of the art', and an 'inventive step' may not be 'obvious to the person skilled in the art'.[21] There is some concern that standards are inadequate, so that patent protection is granted too often over basic genetic material, subtle manipulations of which can be economically very rewarding.[22] More generally, it is argued that what is considered the *proper* scope of patent law is expanding, with property rights moving 'quite far in the direction of private reward over public access'.[23]

The European Patent Convention (EPC), an intergovernmental treaty with a membership extending beyond the European Union, establishes the European Patent Office (EPO), which acts as a centralised office for the granting of patents in Europe.[24] The EPO grants a bundle of national patents, rather

17 Case C-377/98 *Netherlands v European Parliament* [2001] ECR I-7079, para. 27. One of the key objectives set out in the Directive is harmonisation.

18 See Grubb, above n. 8, for examples, particularly Chapter 13 on biotechnology.

19 See for example the discussion of the Nuffield Council on Bioethics, *The Ethics of Patenting DNA: A Discussion Paper* (Nuffield Council on Bioethics, 2002).

20 Llewellyn and Adcock, above n. 8.

21 European Patent Convention (EPC), Articles 54, 55 and 56.

22 Nuffield Council on Bioethics, above n. 19, expresses concern that patents of 'doubtful validity' are granted over DNA sequences. See also the review of the literature on 'bad patents' (generally) in Chris Dent, 'Decision-making and Quality in Patents: An Exploration' (2006) 28 *European Intellectual Property Review* 381.

23 Susan Sell, 'Intellectual Property and Public Policy in Historical Perspective: Contestation and Settlement' (2004) 38 *Loyola of Los Angeles Law Review* 267, p. 320. Sell argues that the balance should be redressed, and also that 'history teaches us that it might'. See also Peter Drahos, 'Biotechnology Patents, Markets and Morality' (1999) 21 *European Intellectual Property Review* 441; Dutfield, above n. 2; James Boyle, 'The Second Enclosure Movement and the Construction of the Public Domain' (2003) 66 *Law and Contemporary Problems* 33.

24 Decisions of the EPO are not binding on the EU's Member States (or the ECJ), but there is a conscious effort at consistency.

than a single pan-European right, which means that matters of infringement, enforcement and revocation (subject to a period of 'opposition proceedings' at EPO level, discussed below) remain matters of national law. There is also EC law specifically on the subject of biotechnology patents, the Biotechnology Directive,[25] and, whilst the Convention has not been amended, the EPO has altered its interpretive rules to achieve consistency with the Directive. The Biotechnology Directive requires Member States to 'protect biotechnological inventions under national patent law',[26] putting beyond doubt the patentability of biotechnological inventions. The Biotechnology Directive was a notoriously contentious piece of legislation, reflecting the 'turbulent state of the public debate on biotechnology'[27] in a decade-long to-ing and fro-ing between European Parliament, Commission and Council. There was fundamental disagreement between the institutions on the appropriate balance between incentives for industry and social and ethical protection. The highly technical and narrow basis of the early drafts vividly demonstrates the failure of the EU institutions at this time to appreciate public offence at developments in biotechnology, and the European Parliament's refusal to adopt the Commission's draft of the Biotechnology Directive is often cited as evidence of a sceptical approach to biotechnology in that institution. The final version of the Directive is more sensitive to social and ethical concerns, and includes the '*ordre public*/morality' exception discussed below.

Internationally, the Agreement on Trade Related Aspects of Intellectual Property Rights (the TRIPS Agreement) was completed in 1995. TRIPS requires patents to be made available for 'any inventions . . . in all fields of technology, provided that they are new, involve an inventive step and are capable of industrial application'.[28] All members of the World Trade Organization (WTO) must sign up to TRIPS, meaning that entry to the world trading system is more or less dependent on patent protection, and TRIPS is backed up by the WTO's powerful dispute settlement procedure.[29] Patent rights must be available under TRIPS without discrimination as to nationality or whether a product is made locally or imported. The conditions for the award of a compulsory licence are also limited. 'Limited exceptions' to the rights of a patentee are

---

25 Directive 1998/44, above n. 6. For detailed discussion, see Denis Schertenleib, 'The Patentability and Protection of Living Organisms in the European Union' (2004) 26 *European Intellectual Property Review* 203; Llewellyn and Adcock, above n. 8, Chapter 7.

26 Directive 1998/44, above n. 6, Article 1.

27 *Howard Florey/Relaxin* [1995] EPOR 541, para. 6.4.4.

28 Article 25(1).

29 Discussed in the context of intellectual property by Gregory Shaffer, 'Recognizing Public Goods in WTO Dispute Settlement: Who Participates? Who Decides?' (2004) 7 *Journal of International Economic Law* 459. See also Chapter 6.

however permitted, 'provided that such exceptions do not unreasonably conflict with a normal exploitation of the patent and do not unreasonably prejudice the legitimate interests of the patent owner, taking account of the legitimate interests of third parties'.[30] This is the provision under which efforts to balance the rights of patent holders and the rights of farmers or researchers, discussed further below, can be considered. TRIPS has been most controversial in respect of pharmaceuticals, which in many developing countries had previously enjoyed little or no intellectual property protection. It is also, however, proving controversial in the context of agriculture, because of the same anxiety about restricting access to vital commodities.

GLOBAL DISTRIBUTION (1): THE TRAGEDY OF THE ANTICOMMONS?

The 'tragedy of the anticommons' provides a vivid metaphor for critics of patent rights in biotechnology, the 'mirror image' of Garrett Hardin's famous tale of the 'Tragedy of the Commons'.[31] By contrast with a commons, in which multiple users are endowed with use rights and none has an effective right to exclude, Michael Heller argues that, in an anticommons, multiple parties are endowed with rights to exclude, but none has an effective use right.[32] Whether anticommons property (rather than just *private* property, which might include the privatisation and ownership of formerly common resources) exists at all in the context of biotechnology patents is of course subject to debate. The under-use of resources in the 'anticommons', however, matches the over-use associated with the commons, and certain problems of potential under-use do arise when we examine the patenting of biotechnological inventions.

Most obviously, the monopoly prices associated with patents and the obligation on farmers to purchase seed every season (rather than save and exchange seed) could make products and technologies unaffordable, particularly in developing countries. If the products of agricultural biotechnology are very

---

30 Article 30.

31 Michael A. Heller, 'The Tragedy of the Anticommons: Property in Transition from Marx to Markets' (1998) 111 *Harvard Law Review* 621, p. 622. See Garrett Hardin, 'The Tragedy of the Commons' (1968) 162 *Science* 1243.

32 See Heller, above n. 31; Michael Heller and Rebecca Eisenberg, 'Can Patents Prevent Innovation?' (1998) 280 *Science* 698. By contrast, see Peter Lee, 'Patents, Paradigm Shifts, and Progress in Biomedical Science' (2004) 114 *The Yale Law Journal* 659, arguing that by closing down 'normal science' patents can actually encourage 'paradigm shifts'.

beneficial or monopolise the market, patents could create a further disadvantage for poor farmers, which should prompt thought about the distributional impacts of agricultural biotechnology. As discussed in Chapters 2 and 3, it is difficult to find a regulatory home for the distributional impacts of new technologies. If patent law exacerbates those distributional impacts, it needs also to consider their regulation.

Beyond the affordability of patented products, patents could restrict ongoing development of that protected product (for example, a seed). Debate about the potential for patent rights to close down research has been particularly lively in respect of investigations into human disease. Essentially, the concern is that data, material or techniques crucial to future research will be patented, perhaps before the full potential of the patented subject is understood.[33] The commercial development of subsequent discoveries becomes extremely costly and difficult because researchers have to identify patents in their field, and then either avoid using those patented processes or materials or negotiate and pay for licences for their use. Whether and to what degree patent rights stimulate or stifle innovation is empirically uncertain,[34] but the potential for patent rights to restrict innovation is generally accepted, if only as a price worth paying for the benefits of patents.[35] As with human health, the patent protection of products and methodologies in agricultural biotechnology could inhibit further research. This is a general concern, but raises particular distributional issues around the investigation of highly localised needs in respect of disease, pests or climate. Locally specific development may not be profitable, but would in the absence of a patent be possible through gradual on-farm or community improvements of crops, as well as by focused public or voluntary sector research. This is again particularly pertinent for developing countries, as it is very likely that the agricultural biotechnology industry will continue to concentrate its research efforts on crops suitable for rich farmers in the developed world. And the risk is that,

---

33 One of the key issues has been a robust application of the 'capable of industrial application' or 'usefulness' criteria. See Rob J. Aerts, 'The Industrial Applicability and Utility Requirements for the Patenting of Genomic Inventions: A Comparison Between European and US Law' (2004) 26 *European Intellectual Property Review* 349.

34 See the discussion in E. Richard Gold, 'Biomedical Patents and Ethics: A Canadian Solution' (2000) 45 *McGill Law Journal* 413; Rebbeca Eisenberg, 'Patents and the Progress of Science: Exclusive Rights and Experimental Use' (1989) 56 *University of Chicago Law Review* 1017. This debate is as old as the debate over the justifications of patent law; see Machlup and Penrose, above n. 15.

35 So for example, in a discussion of the exclusion from patentability of fundamental scientific principles, the US Supreme Court observed that patents 'can discourage research by impeding the free exchange of information', *Laboratory Corporation of America Holdings v Metabolite Laboratories Inc* 126 S Ct 29121.

if patented GMOs come to dominate farming, there is ever less space for local needs.

The existence in many jurisdictions of an 'experimental use' or 'research' defence to patent infringement attempts to provide a direct solution to the possibility that patent rights restrict rather than enhance innovation.[36] The scope of this defence is far from clear even in single jurisdictions, and tends to be applied differently in different jurisdictions, even in Europe. Whilst it could go some way towards preventing the closing down of research in a patented product, it does not currently provide a secure solution to the dilemmas faced here. Most jurisdictions attempt to draw a line somewhere between 'pure' research and commercial exploitation,[37] allowing the defence only if there is no commercial purpose to the research. The boundary between 'pure' and commercial research is however increasingly blurred, restricting the scope of the defence if narrowly applied. But the mere existence of the defence at least recognises that there is a balance to be drawn between the different effects of patents, and hence provides potential to redraw that balance. This, however, simply leads to the fundamental question of what level of monopoly and what level of incentive is appropriate in patent law. The reach of the 'experimental use' defence is not a simple technical detail of patent law, but an enormously complex and deeply political question. A clear identification of the competing justices, in particular a consideration of the very precise problem of important localised (or otherwise uneconomic) improvements, would be a start.

Even a generous approach to an experimental/research defence would be problematic in situations of gradual improvement through use and experience of farmers and communities. The 'farmers' privilege' defence is designed

---

[36] See for discussion William Cornish, 'Experimental Use of Patented Inventions in the EC States' (1998) 29 *International Review of Industrial Property and Copyright Law* 735; Eisenberg, above n. 34. The experimental use defence is not found in the European Patent Convention, although it has been adopted by most Members, drawing on the draft Community Patent Convention.

[37] The English Court of Appeal in *Monsanto v Stauffer* [1985] RPC 515 held that in principle field trials fall within the experimental use exception (which is not limited to laboratory work), but that a distinction is drawn between genuine experimentation and 'amassing statistics to further the commercial exploitation', p. 538 *per* Dillon LJ; seeking to find out something new (including finding out whether a product works in different conditions) can be regarded as an experiment, but trials to demonstrate to a third party that a product works are not. See also *Auchinloss v Agricultural & Veterinary Supplies Ltd* [1999] RPC 397, holding that the provision of samples to a regulatory authority for official approval and not to discover something unknown or to test a hypothesis does not fall within the experimental use exception. The common law experimental use exemption is particularly narrowly drawn in the United States; see Lee, above n. 32.

precisely for this sort of situation. It has its origins in plant varieties legislation, which has traditionally balanced social needs and commercial protection rather differently from patent law. It was incorporated into the Biotechnology Directive, which itself epitomises the greater use of patent protection for plants in the wake of the development of agricultural biotechnology. The farmers' privilege under the Biotechnology Directive applies when a farmer uses the product of a harvest for propagation or multiplication on his or her own holding.[38] It applies only in respect of listed crops, and anyone not falling within the definition of a 'small farmer'[39] must pay 'an equitable remuneration'.[40] The farmers' privilege is specifically designed to safeguard the traditional practice of farmers using seed saved from previous harvests to resow crops. Its existence provides some recognition of the long-term investment of farmers and communities in the genetic resources available to the biotech industry, and indeed recognition of future improvements:

> the past, present and future contributions of farmers in all regions of the world, particularly those in centres of origin and diversity, in conserving, improving and making available these resources, is the basis of Farmers' Rights.[41]

The farmers' privilege allows for on-farm local adaptation of patented crops, or even (although not under the Biotechnology Directive, where it applies only to on-farm seed saving) traditional exchange between farmers. Limits on experimentation continue to apply when we move into the laboratory, or to the seed merchant. Notwithstanding its increasing significance in an age of biotechnology, however, consensus on the desirability of farmers' rights seems to be diminishing. From 1991, the international agreement on plant variety rights has made the farmers' privilege optional for members even in respect of plant variety protection.[42] And the European Commission supports

---

38 Directive 1998/44, above n. 6, Article 11, incorporates Article 14 of Regulation 2100/1994 above n. 12. Note that there is also a 'breeders' privilege', providing similar protection for plant breeders, in the plant varieties legislation. For a critical view of 'farmers' privilege' see Jeremy P. Oczek, 'In the Aftermath of the "Terminator" Technology Controversy: Intellectual Property Protections for Genetically Engineered Seeds and the Right to Save and Replant Seed' (2000) 41 *Boston College Law Review* 627.

39 Defined by reference to the farm's production potential, Regulation 2100/1994 above n. 12, Article 14.

40 This has to be 'sensibly lower than the amount charged for the licensed production of propagating material of the same variety in the same area', Regulation 2100/94 above n. 12.

41 Recitals to *International Treaty on Plant Genetic Resources for Food and Agriculture* (Food and Agriculture Organization, 2001).

42 International Union for the Protection of New Varieties of Plants (UPOV).

corporate farming's resistance to farmers' privilege, 'where farming has become a commercial and *quasi*-industrial activity performed by a small minority of the population and where plant breeding has become an industrial plant breeder's activity'.[43] This fails to acknowledge how the downgrading of the privilege may encourage industrial and capital-intensive farming, as well as potentially reducing the diversity of crops in the EU, by tying every farmer into purchasing seed on the market each season. Farmers' privilege is, though, undoubtedly a more urgent issue in the developing world, where saving and exchanging seeds is closely related to food security.

GLOBAL DISTRIBUTION (2): 'BIOPIRACY' AND BENEFIT SHARING

A further question of global distribution revolves around the legitimacy of the commercialisation by northern-based transnational corporations of products that rely upon genetic resources and/or traditional knowledge indigenous to the developing world. This is a topic of heated and sometimes acrimonious debate, with allegations of 'biopiracy' and neocolonialism made against northern governments and corporations.[44] Essentially, what happens is that industrial or governmental organisations from the developed world turn to the greater biodiversity of the developing world to seek out the raw genetic material for agricultural or pharmaceutical development. Very often, generations of knowledge about the therapeutic or agricultural uses of natural materials, or generations of breeding and use of crops, add value to what looks like 'natural', unworked matter. This is not new of course, as witnessed by the spread of economically useful crops, from potatoes to rubber, around the world over the past several hundred years. But now, rather than the application of blatant colonial military strength, genetic resources are subjected to scientific intervention and their fruits are commodified through intellectual prop-

43 *Communication from the European Communities and their Member States on the Review of Article 27.3(b) of the TRIPS Agreement, and the Relationship between the TRIPS Agreement and the Convention on Biological Diversity (CBD) and the Protection of Traditional Knowledge and Folklore,* IP/C/W/383, 17 October 2002, para. 87. Cited in Llewellyn and Adcock, above n. 8, pp. 192–3.

44 Vandana Shiva, *Protect or Plunder? Understanding Intellectual Property Rights* (Zed Books, 2001), provides a polemical introduction to the issue. See Chris Hamilton, 'Biodiversity, Biopiracy and Benefits: What Allegations of Biopiracy Tell Us about Intellectual Property' (2006) 3 *Developing World Bioethics* 1471, for a discussion of the development of the term 'biopiracy'. See Bronwyn Parry, *Trading the Genome: Investigating the Commodifcation of Bio-Information* (Columbia University Press, 2004) for detailed examination of the issues.

erty law. Issues of fairness in agricultural biotechnology, because of this very direct and immediate link between the corporate property and the genetic or intellectual resources of the developing world, go beyond the usual questions of redistribution between north and south or even the history of colonialism. The 'bio-prospector' may in some cases have paid a fee for the extraction of, say, a plant, as they would if they were extracting coal or gold. But the nature of the resource, the value of which is as information capable of replication far beyond its origin, means that the industry visits only once, rather than returning for more (as it would for coal or gold). The fee to extract the plant can look rather small as a consequence.[45]

In part, the potential unfairness of patenting genetic resources is simply a question of the improper grant of patents, to which a partial solution may be more robust and critical investigation of applications. There are cases in which patents have been granted in spite of existing indigenous use or knowledge of the 'invention'. Turmeric and neem are perhaps the most notorious examples. A patent was granted in the US in 1995 for the use of turmeric in healing wounds. On re-examination, however, all claims were cancelled following citation of 'prior art', consisting of traditional knowledge and use of the healing properties of turmeric in India. Similarly, in 2000 the EPO revoked a patent granted in respect of the use as a fungicide of extracts from the seed of the neem tree.[46] Again, the neem tree had been widely used for centuries as a fungicide in India. The 'invention' in these sorts of cases should be unpatentable in normal patent law on the basis of the 'prior state of the art'. However, when a patent is claimed, the prior state of the art is generally considered by reference to (at best) published scientific information. Traditional knowledge will often not be unearthed by this route, although there are proposals to at least reduce this problem by creating a database of traditional knowledge.[47] Whilst a patent can be

45 On traditional knowledge, see for example Michael Blakeney, 'The Protection of Traditional Knowledge under Intellectual Property Law' (2000) 22 *European Intellectual Property Review* 251; Peter Drahos, *Towards an International Framework for the Protection of Traditional Group Knowledge and Practice* (UNCTAD–Commonwealth Secretariat Workshop on Elements of National Sui Generis Systems for the Preservation, Protection and Promotion of Traditional Knowledge, Innovations and Practices and Options for an International Framework, 2004); Graham Dutfield, 'The Public and Private Domains: Intellectual Property Rights in Traditional Knowledge' (2000) 21 *Science Communication* 274; Parry, above n. 44.

46 Confirmed by the Technical Board of Appeal in 2005.

47 Various documents produced by the TRIPS Secretariat outline the issues and the difficulties; see http://www.wto.org/english/tratop_e/trips_e/art27_3b_e.htm (accessed December 2007). This requires at least awareness of the traditional knowledge, and its amenability to written description, and may not always be culturally or practically appropriate.

challenged, those wishing to bring such a challenge face enormous and obvious information and resource barriers. The Indian Government was involved in the challenge to the turmeric patent, a group of non-governmental organisations in the case of neem.

Moreover, even 'good' patents, in respect of a genuinely inventive step, if they build on a genetic resource or traditional knowledge, raise questions of equity. It is often pointed out that such a patent will not lead to the grant of a general monopoly that denies access to the underlying genetic resource or traditional knowledge.[48] Similarly, patents are not available on the crops that have been grown for millennia, so, in theory, farmers will only use patented crops if the extra costs are worthwhile. There are nevertheless profound and complex challenges involving divided understandings of property, knowledge and development. At its very simplest, there is a problem of inadequate recompense and inadequate sharing of benefits. Without intervention, genetic resources and traditional knowledge come more or less free. Traditional knowledge is of course diverse in its manifestations, but because it is often public and collective, or has evolved over long periods of time, it is not subject to patent rights (and so ownership), the framework of which is designed with the norms of western science in mind.[49] And genetic resources have traditionally been considered part of the 'common heritage of human kind', and freely available.[50] There is a danger that what leaves the developing world as 'common heritage' is returned as a commodity.[51] It should not be thought that issues of appropriation without consent are of concern only to the developing world. So to take an example discussed in Chapter 2, the commercialisation of the *Bacillus thuringienesis* (Bt) gene in GM plants appropriates for short-term commercial benefit the pesticide effects of a microbe used sparingly by the

---

48 Grubb, above n. 8; Hanns Ullrich, 'Traditional Kowledge, Biodiversity, Benefit-Sharing and the Patent System: Romantics v. Economics?' in Francesco Francioni and Tullio Scovazzi (eds), *Biotechnology and International Law* (Hart Publishing, 2006); Christopher Heath and Sabine Weidlich, 'Intellectual Property: Suitable for Protecting Traditional Medicine?' (2003) *Intellectual Property Quarterly* 69.

49 See the sources cited above n. 45; also N.S. Gopalakrishnan, 'TRIPS and Protection of Traditional Knowledge of Genetic Resources: New Challenges to the Patents System' (2005) 27 *European Intellectual Property Review* 11. Some see the patent system as designed more specifically with the US in mind; see William Kingston, 'Why Harmonisation is a Trojan Horse' (2004) 26 *European Intellectual Property Review* 447.

50 Mary Footer, 'A Tale of Two Commons: Plant Genetic Resources and Agricultural Trade Reform' in Han Somsen (ed.), *The Regulatory Challenge of Biotechnology: Human Genetics, Food and Patents* (Edward Elgar, 2007).

51 Hamilton, above n. 44.

organic movement.[52] But there has been no opportunity to discuss the relative social benefits from the use of Bt in different types of farming and different crops, or the social value of a non-resistant insect population.

The issue of benefit sharing is now widely discussed at all levels, most importantly perhaps within the WTO (although there are no signs of imminent progress),[53] and acknowledged, to some degree at least, in a number of legal instruments. The 1992 Rio Convention on Biological Diversity, for example, recognises national sovereignty over natural, including genetic, resources,[54] and provides for access to such resources 'on mutually agreed terms', and 'subject to prior informed consent'.[55] One of the objectives of the Convention is 'the fair and equitable sharing of the benefits arising out of the utilization of genetic resources'.[56] The Convention is also concerned with the 'knowledge, innovations and practices of indigenous and local communities embodying traditional lifestyles', referring to the 'equitable sharing of the benefits arising from the utilization of such knowledge, innovations and practices'.[57] It has been suggested that linking the benefit-sharing elements of the Convention on Biological Diversity (and benefit sharing might include technology transfer or local employment as well as cash) to eligibility for a patent would sharpen protection. An obligation to provide evidence in a patent application of prior informed consent from the holders of the traditional knowledge or genetic resources would similarly give the Convention teeth. An obligation to disclose the source of the traditional knowledge or genetic resources on which an invention is based could help both to avoid the problems of patenting an 'invention' that relies on traditional knowledge and to make 'the fair and equitable sharing of benefits' easier. There are several different ways to put such

52 See Richard Hindmarsh, Geoffrey Lawrence and Janet Norton, 'Bio-utopia: The Way Forward?', and Anna Salleh, 'Wearing out our Genes? The Case of Transgenic Cotton', both in Hindmarsh, Lawrence and Norton (eds), *Altered Genes; Reconstructing Nature: The Debate* (Allen and Unwin, 1998).

53 TRIPS contains an obligation to review the operation of Article 27.3(b), which permits members to exclude plants and animals from patentability, if they provide 'an effective sui generis system' for the protection of plant varieties. For a discussion of the positions of different members of the WTO, see the summaries prepared by the Secretariat to the Council for TRIPS: *The Relationship Between the TRIPS Agreement and the Convention on Biological Diversity* IP/C/W/368/Rev.1; *Review of the Provisions of Article 27.3(b)* IP/C/W/369/Rev.1; *The Protection of Traditional Knowledge and Folklore* IP/C/W/370/Rev.1.

54 Articles 3 and 15. Article 2 provides definitions: 'Genetic resources' means genetic material of actual or potential value; 'Genetic material' means any material of plant, animal, microbial or other origin containing functional units of heredity.

55 Article 15.4 and 15.5.

56 Article 1.

57 Article 8j.

an obligation in place, including mandatory or voluntary disclosure, at the national or international level.[58] Sanctions for failure to disclose could be placed within the patent system, for example by making the validity of the patent dependent on disclosure, or outside the patent system, for example through damages.[59] More modestly, rather than looking to intellectual property law, a contractual solution is sometimes put forward to the distributional impact of the patent system, a form of self-regulation. It is not unusual to see industries concerned about their reputation entering into agreements with developing countries to provide for longer-term benefits, including in some cases royalty payments on any finished product.[60]

The Convention on Biological Diversity by no means dictates any particular solution to the dilemmas inherent in the use of genetic resources and traditional knowledge for commercial gain. It is far from self-explanatory in its key terms (for example, the meaning of both 'fair and equitable' and 'benefits' is unclear, as is the type of 'utilization' that is envisaged). And, whilst it provides real impetus towards benefit sharing, the Convention also includes the proviso that 'all rights over those resources and to technologies' should be taken into account. The relationship between the Convention and TRIPS is not clear,[61] and patent rights over products derived from traditional knowledge or genetic resources continue to be granted in the normal way. When the Netherlands challenged the validity of the Biotechnology Directive, it argued that the patentability of biotechnology inventions breaches the principle of equitable sharing in the Convention on Biological Diversity. The ECJ concluded that nothing in the Convention demands that 'conditions for the grant of a patent for biotechnological inventions should include the consideration of the interests of the country from which the genetic resource originates or the existence of measures for transferring technology'.[62] We should also bear in mind that

---

58 The desirability of disclosure is apparent in Directive 1998/44, above n. 6, Recital 27, providing that the 'geographical origin' of an invention based on or using biological material of plant or animal origin or 'where appropriate' and 'if known'. This exhortation is however absent from the operative Articles of the Directive, and is 'without prejudice to the processing of patent applications or the validity of rights arising from granted patents'.

59 For detailed discussion, see Secretariat to the Council for TRIPS: *The Relationship between the TRIPS Agreement and the Convention on Biological Diversity* IP/C/W/368/Rev.1. Note the observation of Jon Santamauro, 'Reducing the Rhetoric: Reconsidering the Relationship of the TRIPS Agreement, CBD and Proposed New Patent Disclosure Requirements Relating to Genetic Resources and Traditional Knowledge' (2007) 29 *European Intellectual Property Review* 91, that the source of the material was disclosed in the patent applications for turmeric and neem.

60 See Parry, above n. 44.

61 IP/C/W/368/Rev.1, above n. 53.

62 *Netherlands*, above n. 17, para. 66.

the United States, a technology-wealthy nation, is not a party to the Convention on Biological Diversity.

The identification of interests, if not rights, in natural resources and indigenous knowledge, which are deserving of recompense, could be a response to certain distributional dilemmas of patenting. It is not however straightforwardly welcome. The reconceptualisation of biological diversity as a 'genetic resource' renders such genetic resources more amenable to property-type analysis.[63] As such, it is a further 'enclosure of the commons',[64] the latest phase in the privatisation of common property, and successor to the enclosures of common land in Europe from the 16th century. Although it is a response to 'enclosure' through patent rights, as with the earlier enclosures, views differ as to whether the enclosure contributes to the public good through economic efficiency and increased production or redistributes from poor to rich, dislocating communities and environments.[65] Property in traditional knowledge also raises considerable difficulties, for example as to who holds it – traditional knowledge and use of the neem tree, for instance, involved many thousands of people in a number of countries – before even considering whether marketisation and economic reward are appropriate to the many different social contexts of traditional knowledge.[66] It is unlikely that a reliance on consent or compensation will be able to avoid excluding some groups or individuals. Whatever the difficulties, however, genetic resources and traditional knowledge are major resources available to developing countries, whilst developed countries and their corporate entities hold the technological resources to manipulate that material for commercial ends. Given the uneven distribution of benefits, some sort of intervention seems appropriate.

There are no straightforward solutions to the distributional impacts of biotechnology. Some simply reject the applicability of the patent system (and indeed the market system) to these types of resources.[67] Recompense and consent (mandatory or self-regulatory) both require the identification of parties to agreement, which might include state or local government, communities or individuals. The Convention on Biological Diversity looks to national sovereignty over genetic resources, which is straightforward for the bioprospector, but assumes the responsiveness of government to community.

---

63 Hamilton, above n. 44.

64 Footer, above n. 50.

65 See especially the discussion in Boyle, above n. 23. Boyle argues that, whilst the impact of the first enclosure remains disputed, even if it was positive there is no reason to think that such advantages will be transferred to the intellectual property arena.

66 Ullrich, above n. 48; Heath and Weidlich, above n. 48.

67 See for example Shiva, above n. 44.

There are similar practical and conceptual challenges to increasing disclosure through the patenting system. Most genetic resources have many possible origins, which raises questions as to the disclosure of one potential source rather than another. The difficulty of tracing finished products back to particular sources in particular countries means, moreover, that longer-term benefits are largely at the discretion of the patent holder.[68] And further, whilst the privatisation and monopoly implied by patents is perhaps least attractive in respect of a staple food crop suitable for growth by poor farmers in the developing world, this is the area in which we are most critical of the biotechnology industry for failure to invest in socially valuable research. The Biotechnology Directive points to the positive potential of the incentive provided by patents in this area, asserting the importance of biotechnology to developing countries, 'both in the field of health and combating major epidemics and endemic diseases and in that of combating hunger in the world', and proposing the use of the patent system 'to encourage research in these fields'.[69] It seems unlikely that the patent system alone will encourage such research, and research is currently focused primarily on the needs of those best able to pay. This is well recognised in the pharmaceutical context by the notion of 'orphan' diseases, research into which is not incentivised by the patent system.[70] Perhaps in recognition of this, Recital 18 of the Biotechnology Directive refers to the need for technology transfer to developing countries. Again, however, this is a topic that seems to demand more than reliance on markets. Watching the pharmaceutical multinationals (for good economic reasons) more or less ignore the medical needs of the world's poor does not inspire faith in the ability of a commercially dominated agricultural industry to feed the world.

Rather than supporting any particular approach to these difficulties, the purpose of this brief review has been to identify how patent law highlights (and even exacerbates) some of the concerns about GMOs discussed in Chapter 2. It is also clear that patent law has a role regulating (even if only implicitly) some of the impacts of biotechnology. The north/south distributional impacts of biotechnology are extremely complex and deserve more discussion than it is possible to provide here. Distributional questions should not be forgotten in the colonisation of regulation by the risk framework, and we should note that patents provide an alternative forum of regulation.

The spread of European/North American standards of patent protection through the TRIPS Agreement, described as 'harmonisation with a vengeance'

---

68 Parry, above n. 44.
69 Directive 1998/44, above n. 6, Recital 11.
70 Directive 1998/44, above n. 6, Recital 18.

for its aggressive prioritisation of commercial interests,[71] makes the distributional impact of patent law an urgent matter. TRIPS positions intellectual property rights as a trade issue rather than a social, ethical or environmental issue. Notwithstanding the provision of a 'morality' clause in some patenting regimes (including TRIPS), as discussed below, this characterisation of intellectual property reflects the dominance of the economic purpose of intellectual property, and also the perspective of the private sector advocates of the Agreement.[72] More generally, it will be recalled that concern about the dominance of agriculture by large corporations is a persistent theme for those who resist biotechnology. It has been the case for some years that patents in the biotechnology industry are concentrated in relatively few hands.[73] The transnational corporation has very specific interests, and its research, development and marketing activities are likely to be directed towards satisfying those interests.

LOCAL RELATIONSHIPS: MORE COEXISTENCE PROBLEMS

The previous chapter discussed some of the challenges that widespread GM farming in the EU might be expected to pose for other farmers. Patent law further determines the nature of the local and regional impact of GM farming. Most generally, the monopoly rights granted to patent holders can be expected to raise costs for farmers, and to interrupt well-established farming practices, including saving seed for resowing. More specifically, the possibility of 'innocent infringement' of a patent, if non-GM farmers' crops become 'contaminated' by a neighbour's GM crop, has the potential radically to change relationships between neighbours. Because patent infringement is a matter of strict liability, the 'innocent infringer' is highly vulnerable to legal action by the holder of the patent, in respect of unsought and unwanted GM presence in their crops. And patent holders do uncompromisingly defend their patent

---

71 Kingston, above n. 49, p. 455.

72 For detailed discussion, see Susan K. Sell, *Private Power, Public Law: The Globalization of Intellectual Property Rights* (Cambridge University Press, 2003); Dutfield, above n. 2.

73 See for example David Schimmelpfennig and John King, 'Ag Biotech Patents on the Move' (2005), available at http://www.ers.usda.gov/AmberWaves/June05/Findings/AgBiotechPatents.htm (accessed December 2007); also Jennifer Clapp, 'Transnational Corporate Interests and Global Environmental Governance: Negotiating Rules for Agricultural Biotechnology and Chemicals' (2003) 12 *Environmental Politics* 1, p. 13.

rights, as the well-known story of the Canadian litigation in *Monsanto v Schmeiser* demonstrates.[74]

This case arose out of an action brought by Monsanto against a Saskatchewan farmer for infringement of its patent for GM canola. Monsanto's patent was for genetically engineered genes, and cells containing those genes, which when inserted in plants make them resistant to glyphosate herbicides, including Monsanto's 'Roundup'. Monsanto had sold and licensed the use of 'Roundup Ready' canola to a number of farmers in Schmeiser's area. Although he never purchased the GM canola, by 1998, 95 to 98 per cent of Schmeiser's crop was composed of such plants. He claimed that the Roundup Ready plants came from adventitious presence (cross-pollination, volunteers, spills of seeds) in previous generations of seed, saved for the 1998 crop. The trial judge found that 'none of the suggested sources could reasonably explain the concentration or extent' of the presence.[75]

The first question in the litigation was the validity of Monsanto's patent. This was especially interesting because, between the Court of Appeal and Supreme Court decisions, the Supreme Court had handed down its *Oncomouse (Canada)* decision. This decision is discussed further below, but in essence it concluded that higher life forms, including mice and plants, are not patentable in Canada.[76] However, Monsanto's patent was over the genes and modified cells that make up the plant, rather than the plant itself, and on that basis the Supreme Court in *Schmeiser* found it to be valid. The *Oncomouse (Canada)* decision had knock-on effects, though, on the key question of whether Mr Schmeiser had 'used' the patented invention, in this case the cell or gene (not the plant). The Court had to decide whether, by cultivating plants containing the cell and gene, the defendant used the patented components. The majority held that he had. Both *Schmeiser* and the *Oncomouse (Canada)* decision were reached by a narrow majority, and the change of approach, albeit not overruling the bar on patenting of higher life forms, may indicate a changing balance of the Court.[77] The minority in *Schmeiser* took the position that the majority approach renders the restriction on patenting of higher life forms 'meaningless'.[78] The minority would have resolved the question of 'use' by reference to the scope of Monsanto's patent claim (not, of necessity, a claim to the plant itself), and would have prevented the protection of the plant itself. When

---

74 *Monsanto Canada v Schmeiser* [2004] 1 SCR 902. For an excellent review, see Robert Burrell and Stephen Hubicki, 'Patent Liability and Genetic Drift' (2005) 7 *Environmental Law Review* 278.

75 *Monsanto*, above n. 74, para. 6.

76 *Harvard College v Canada (Commissioner of Patents)* [2002] 4 SCR 45.

77 Burrell and Hubicki, above n. 74

78 *Monsanto*, above n. 74, para. 167.

considering the implications of *Schmeiser* in Europe, though, it is important to note that the non-patentability of higher life forms was a central element of the dissent, suggesting that a farmer in the EU will have even more of an uphill struggle escaping liability.

More generally on the defendant's infringing 'use' of the patented cells, the Court held that there is a presumption that a defendant in possession of patented material has indeed used it. This presumption can be rebutted 'by bringing credible evidence that the invention was neither used, nor intended to be used, even by exploiting its stand-by utility'.[79] The defendant had argued that the scope of the patent should be restricted to the way in which the inventor intended the invention to be used, that is, in conjunction with Roundup or some similar herbicide. Since he had not sprayed his fields, it was argued that Schmeiser's activities were non-infringing. The Court set great store by the 'stand-by utility' of the properties of the patented genes and cells, whereby the farmer can spray in the future if he chooses to do so, or even benefit by selling seed on in the future.[80] Whether the patented invention was in fact of any assistance to the defendant, or economically beneficial, was not relevant.

However, the lack of benefit to Mr Schmeiser was, on the specific litigation history of this case, relevant to remedy. Monsanto had sought not damages but an account of profits (an alternative open to them under the Canadian Patents Act). The Supreme Court found that the defendant's profits were exactly what they would have been had he planted non-GM canola. And, given that the Roundup was not sprayed, no agricultural advantage had been gained. The Court dealt fairly swiftly with these difficult points, and it may be that the controversial nature of the substantive decision on infringement was making its influence felt here.[81] This is however anything but a pyrrhic victory for Monsanto. Monsanto is notoriously diligent in its pursuit of patent infringers, and regardless of the absence of remedies (tied in any event to the pleadings in the individual case) the fact of infringement remains.[82] Litigation, moreover, occurs only at the extremes. The larger biotechnology corporations allocate considerable resources to investigating infringement of their patents (Monsanto's hiring of retired Canadian Mounties is often cited, with a combination of amusement and despair), and more common than litigation will be generally invisible demands for the

---

79 *Monsanto*, above n. 74, para. 56.

80 *Monsanto*, above n. 74, para. 85.

81 Burrell and Hubicki, above n. 74.

82 Note that some jurisdictions, including the UK, provide that damages cannot be sought against an infringer able to establish that 'at the date of the infringement he was not aware, and had no reasonable grounds for supposing, that the patent existed', Patents Act 1977, section 62. This does not really cover our 'innocent infringer'.

alleged infringement to cease.[83] Settlement as well as litigation is likely to advantage the powerful and well-resourced, and it has been said that even an invalid patent held by a well-resourced firm is more valuable than a valid one owned by a firm with limited resources to defend it in the courts.[84] The simple fear of litigation presumably has a great influence on research and development activities and on farming practices.

The vulnerability of the wholly 'innocent infringer' to actions for patent infringement in respect of the unsought presence of patented material in his or her crop is real, but the Canadian Supreme Court in *Schmeiser* attempted to provide some protection. The majority explicitly denied that its decision was about the 'innocent discovery by farmers of "blow-by" patented plants on their land or in their cultivated fields'.[85] Whilst knowledge is not a necessary part of a patent infringement, the Court provided that what a person does on becoming aware of the presence of patented material can be used to rebut the presumption of use arising from possession.[86] So, for example, 'showing that they acted quickly to arrange for its removal, and that its concentration was consistent with that to be expected from unsolicited "blow-by" canola' might rebut that presumption of use.[87] Monsanto at least seems to have been convinced that Schmeiser was cynically abusing the presence of the GM seed on his property, and the Court did not see Schmeiser as an innocent infringer, but as a farm company which 'actively cultivated canola containing the patented invention as part of their business operations'.[88] The decision seems to rest not on the 'perhaps adventitious' arrival of the GM canola on the defendant's land: 'What is at stake in this case is sowing and cultivation, which necessarily involves deliberate and careful activity on the part of the farmer'.[89]

The Supreme Court decision mitigated the most worrying elements of the decisions of the lower courts, which had not placed any reliance at all on the knowledge of Mr Schmeiser in their findings of infringement, and demonstrated at least some awareness of the difficulties faced by the innocent infringer. Nevertheless, the decision emphasises certain problems for non-GM

83 Burrell and Hubicki, above n. 74. Center for Food Safety, *Monsanto vs. U.S. Farmers* (Center for Food Safety, 2004) provides a polemical review of litigation and settlements in the US.

84 Kingston, above n. 49, p. 459.

85 *Monsanto*, above n. 74, para. 2.

86 The Supreme Court does not consider in detail other possible ways to infringe, for example whether planting the seed might be 'making' the invention (although 'not inclined' to see as making, para. 26); see Burrell and Hubicki, above n. 74.

87 *Monsanto*, above n. 74, para. 86.

88 *Monsanto*, above n. 74, para. 87.

89 *Monsanto*, above n. 74, para. 92.

farmers. Any farmer who knows (or possibly ought to know) about the presence of the GM material is unable to risk using or saving seed. And the Supreme Court actually said nothing to guarantee that the absence of knowledge in itself will rebut the presumption of use, presumably because the defendant will at least have been informed of the presence by the patent holder. To rebut the presumption of use of the patented material, our innocent infringer needs to '[act] quickly to arrange for its removal'.[90] There is no suggestion of a 'reasonable steps' proviso, or similar. As the dissent pointed out, rebutting the presumption of use 'would likely prove difficult once the innocent infringer became aware that the genetically modified crop was present – or was likely to be present – on his or her land and continued to practice traditional farming methods, such as saving seed'.[91] The entire burden of avoiding patented material is placed on this innocent infringer.

If this approach is followed in the EU, patent rights subvert the intention of the coexistence approaches discussed in Chapter 4. Explicit regulation for coexistence provides that the cost of separating GM, conventional and organic crops will lie with the farmer introducing a new form of production, generally initially the GM producer. As discussed in Chapter 4, the greater interest of the organic farmer in maintaining segregation is likely to mean that in practice the organic sector will not escape the costs of measures such as crop separation, volunteer control and machinery maintenance. But that disruption of farming practices is increased by the provision of remedies to the patent holder if the non-GM farmer does not manage to avoid GM material. And there is every reason to expect the EC legal context to be at least as burdensome as the Canadian for neighbours of GM farmers. Moreover, coexistence measures to be taken by the GM farmer, it will be recalled, only kick in over the 0.9 per cent threshold. Presumably there is potential for patent infringement below that threshold.

The potential for innocent infringement risks putting poorly considered and novel burdens on neighbouring farmers in all jurisidictions. This is compounded by the difficulty (discussed in Chapter 4) of bringing a claim in tort in respect of 'contamination' of crops. There is a real discrepancy between the industry's claim to control over the technology in respect of its own patent rights and a complete denial of control when it comes to liability. This 'unacceptable incongruity'[92] highlights the importance of the post-authorisation framework when thinking about the objections to GMOs. Moreover, whilst the

---

90 *Monsanto*, above n. 74, para. 84.

91 *Monsanto*, above n. 74, para. 159.

92 Martin Phillipson, 'Giving Away the Farm? The Rights and Obligations of Biotechnology Multinationals: Canadian Developments' (2005) 16 *King's College Law Journal* 362, p. 372.

basic fact of strict liability patent infringement does apply in the EU, the application of the more defendant-friendly aspects of the Canadian decision in other jurisdictions remains uncertain.[93]

Alternative remedies for innocent infringers are not easily found. The farmers' privilege, discussed above, will not apply to a case of unwanted presence, but only to replanting of crops initially sown with permission of the patent holder. The 'exhaustion' defence found in the Biotechnology Directive is also unlikely to assist.[94] This applies where the patent owner has placed biological material on the market and where multiplication or propagation of that material necessarily results from the application for which the material was marketed. The problem is when the multiplication or propagation can be said 'necessarily' to result from the purpose for which the material was marketed. The process of making beer, for example, inevitably involves the multiplication of yeast and the exhaustion provision would ensure that the purchaser of the yeast would not infringe the patent by using the yeast in the normal way, without the need to rely on some form of implied licence. By contrast, this provision is not likely to protect a farmer who does not have a pre-existing commercial relationship with the patent owner. In any event, exhaustion does not apply if the material 'is subsequently used for other propagation or multiplication',[95] which arguably describes the situation of the 'innocent infringer' resowing contaminated seeds.

Relationships between neighbours are intensively affected by patent law, potentially in a manner disruptive of the explicit regulation of coexistence. More considered allocation of responsibilities between farmers may well have led to the consideration of an 'innocent infringer' defence, a way to balance the regulatory impact of patent law.

PATENTS ON LIVING ORGANISMS

Whether living organisms are suitable subjects for a patent is now largely resolved in Europe and North America. The convoluted and contentious recent history of such patents, however, demonstrates the intensity of the controversy, and the world-infamous Harvard oncomouse is a useful example. The oncomouse was developed by researchers at Harvard University, with funding from DuPont. An 'oncogene', a gene that promotes cancer, is inserted into a fertilised mouse egg. Through breeding, mice susceptible to cancer (and so useful for research into causes and cures) are produced. The oncomouse was

---

93 Burrell and Hubicki, above n. 74.
94 Directive 1998/44, above n. 6, Article 10.
95 Directive 1998/44, above n. 6, Article 10.

patented with little ado in the US, patented with much ado in Europe, and denied a patent in Canada. We should note that members are allowed under TRIPS to exclude plants and animals (other than micro-organisms) from patentability.[96]

The seminal US decision of *Diamond v Chakrabarty* involved an application for a patent on an artificially produced strain of bacterium, capable of breaking down crude oil.[97] The Supreme Court confirmed, by a narrow (five/four) majority that existing legislation allowed for the patenting of living organisms. According to the inventor, the only reason that a patent over the bacterium as well as the process of creating it was applied for at all was that his employer (General Electric)'s lawyers were more accustomed to dealing with traditional physical inventions like refrigerators and jet engines than biology. They operated in a culture in which patents are available for anything new and useful, rather than a culture more accepting that 'you can't patent life'.[98] The Patent Office's Board of Appeal, however, proceeded from the more conventional assumption that a 'product of nature' was not patentable, and concluded that there was equally a bar on patenting a living organism. The Supreme Court dealt with the issue as a narrow one of statutory interpretation, the majority determining that patenting living organisms would not require explicit legislative provision, identifying the bacterium as a 'product of human ingenuity' rather than a product of nature. *Diamond v Chakrabarty* was subsequently applied to other organisms, and ultimately to the patenting of the oncomouse in the US.[99]

The story of the oncomouse has been more problematic in Canada and Europe. As in the US, the Canadian Supreme Court addressed the case as a narrow question of statutory interpretation, rather than social acceptability.[100]

---

96 Article 27.3(b). Plant varieties must be protectable either by patent or by 'an effective sui generis system', ibid.

97 *Diamond v Chakrabarty* 447 US 303, 100 SCt 2204. See Sheila Jasanoff, *Designs on Nature: Science and Democracy in Europe and the United States* (Princeton University Press, 2005), Chapter 8, for a discussion of the successful 'depoliticisation' of this question through its resolution in the judicial rather than political forum. See also the discussion in Daniel J. Kevles, *A History of Patenting Life in the United States with Comparative Attention to Europe and Canada* (European Group on Ethics in Science and New Technologies to the European Commission, 2002).

98 Kevles, above n. 97, pp. 14–15. The British Patent Office had granted a patent over the bacterium in 1976, Grubb, above n. 8, p. 248.

99 US 4,736,866. The Supreme Court in 2001 confirmed the Patent and Trademark Office's practice of awarding patents for plants in *JEM Ag Supply*, above n. 13.

100 *Harvard College,* above n. 76. The legislation applies to 'inventions', limited to any 'art, process, machine, manufacture or composition of matter', Section 2. Note

This decision, however, does indicate the difficulty of drawing such distinctions, as it rests in part on precisely these questions of social acceptability. A patent for the oncomouse was denied, although patents were granted for the process for genetically modifying a mouse in the laboratory. The narrow (five/four) majority in the Supreme Court concluded that higher life forms, including seeds and plants (and oncomice), are not patentable, although lower life forms (including, *obiter*, the oncomouse egg) are. There are obviously some difficult lines to be drawn in Canadian law, but it is not necessary here to go into the detail of the statutory interpretation. The basic question was whether, in the absence of explicit provision, higher life forms are or are not covered by legislation. As in the US, the concentration on statutory interpretation demonstrates some reluctance to engage in value debate, but there is nevertheless some discussion in both majority and minority judgments of the broader context. In Canada, the fact that the Patent Act 'fails to address many of the unique concerns that are raised by the patenting of higher life forms' must indicate that it was not intended to extend to higher life forms, and is not 'the appropriate vehicle to protect the rights of inventors of this type of subject matter'.[101] When there is 'a policy issue that raises questions of great significance and importance and that would appear to require a dramatic expansion of the traditional patent regime', coverage needs 'explicit legislative direction'.[102] This is precisely the opposite conclusion from the US Supreme Court.

The approach of the EPO has gone beyond legislative interpretation because, by contrast with the Canadian and US systems, the EPC explicitly provides for the consideration of moral issues when deciding whether to grant a patent. The morality exception will be discussed in the next section, but, on the specific question of patenting living organisms, the Harvard oncomouse patent application was initially rejected in 1989, before on appeal it was held that the EPC contains no bar on patenting animals. It was not a foregone conclusion that plants and animals would be patentable under the EPC. Aside from the morality exception, the EPC precludes the patenting of 'plant or animal *varieties*', an exclusion contained also in the Biotechnology Directive.[103] The distinction between plants/animals and plant/animal varieties is not easily drawn, and was initially a major limitation on patentability. However, in *Novartis* a distinction is drawn in such a way as to allow the grant

---

that in the European litigation over the oncomouse, T-315/03 *Harvard/Transgenic animal* [2005] EPOR 31, the Canadian decision is explicitly said to rest on statutory interpretation, and 'quite clearly does *not* establish that animal patents arouse public unease', para. 13.2.16.

101 *Harvard College*, above n. 76, paras 120 and 167, *per* Bastarache J for the majority.

102 *Harvard College*, above n. 76, para. 155, *per* Bastarache J.

103 EPC, Article 53(b); Directive 1998/44, above n. 6, Article 4(1)(a).

of a patent over transgenic plants, indicating the increasing acceptance of patentability in this area.[104] *Oncomouse (Europe)* was remitted to the Examining Division with explicit reference to the 'ordre public'/morality exception to the EPC. Eventually, as will be discussed below, the oncomouse application was assessed against this criterion, and a patent was awarded.

Whilst hugely (and rightly) controversial, the patentability or not of higher life forms does not on its own determine the impact of patent law in this area. A patent might be sought in different aspects of the technology, depending in part on the patent protection available. More straightforward possibilities for GM plants include the actual plant, a plant variety, an isolated gene, plant cells or plant DNA, promoters or markers, and techniques and processes such as cloning or regeneration. So even if a plant is deemed ineligible for patent protection (explicitly allowed under TRIPS), and although there is likely to be lengthy discussion of the meaning of 'plant', there may be a range of ways to protect the invention.[105] The fact that *Monsanto v Schmeiser* arose in a jurisdiction that does not allow the patenting of higher life forms (including plants) reminds us that in practical terms the 'patenting life' decision does not close down the debate. Moreover, even with successful restrictions on patenting life, corporations are likely to turn to alternative means of protection such as terminator technology (discussed in Chapter 2) or trade secrets. It is a consistent theme of the patent debate that denying a patent does not prevent either invention or exploitation, but simply allows anyone to exploit.[106] The complete denial that restrictions on patentability would have any impact, however, rather assumes that patents have no impact, contrary to arguments that they are a crucial stimulus to invention and prosperity. The proper scope of patent law is a genuine dilemma, and will be pursued a little further in the next section.

THE MORALITY PROVISO

Patent law is generally 'technology neutral', so that once the relevant legal criteria are met there is no need to examine further reasons for the grant or denial of a patent.[107] Courts and officials involved in patents have fairly consistently insisted that they cannot provide an appropriate forum for the discussion of ethical or social questions provoked by invention. This 'patent

---

104 See above n. 12.

105 Mark D. Janis, 'Sustainable Agriculture, Patent Rights and Plant Innovation' (2001) 9 *Indiana Journal of Global Studies* 91.

106 And again, this is a longstanding debate; see Machlup and Penrose, above n. 15.

107 Llewellyn and Adcock, above n. 8, pp. 15–16.

community' has every interest in avoiding reference to social and ethical debate, instead emphasising the apparent objectivity of a rational and neutral application of the legal framework. In this, they are no different from other regulators. However, like other regulators, patent law does not operate in a moral vacuum, and does imply the application of value judgments, if silently. Determined resistance to biotechnology patents has forced some of these issues into the public domain.

Morality and public policy are explicit factors for consideration in the grant of a patent in Europe. Article 53(a) EPC provides that 'inventions the publication or exploitation of which would be contrary to "ordre public" or morality' shall not be patented, exceptions to patentability that are, in common with other exceptions, narrowly construed.[108] The equivalent provision in the Biotechnology Directive is that 'Inventions shall be considered unpatentable where their commercial exploitation would be contrary to *ordre public* or morality'.[109]

The basic content of 'ordre public' (translated into English as 'public policy' in Article 30 of the EC Treaty), as interpreted by the EPO, embraces 'the protection of public security and the physical integrity of individuals as part of society'.[110] In the absence of more specific provisions, 'ordre public' extends to the protection of the environment under the EPO, the Biotechnology Directive[111] and also TRIPS.[112] The approach to environmental protection looks rather far-reaching: 'inventions the exploitation of which is likely . . . seriously to prejudice the environment are to be excluded from patentability as being contrary to "ordre public"'.[113] It is, however, rather under-used, not least because the extreme stringency of the EPO's approach, discussed below, makes it difficult to imagine what sort of invention would fail the 'environmental protection' part of ordre public. There is even less room for the consideration of the most difficult kinds of environmental uncertainties in the patent system than there is in the risk regulation system.

108 Case T-356/93 *Plant Genetic Systems /Glutamine synthetase inhibitors*[1995] EPOR 357.

109 Directive 1998/44, above n. 6, Article 6(1). TRIPS also provides the option of an ordre public/morality exception, Article 27(2). Note that the exception does not consider the morality of patenting, or indeed of the invention itself, but the morality of 'exploiting or publishing' (EPC), and 'commercial exploitation' (Directive 1998/44, above n. 6).

110 *Plant Genetic Systems,* above n. 108, para. 5 (Technical Board of Appeal), cited in *Harvard*, above n. 100, para. 10.2.

111 *Netherlands*, above n. 17: 'a genuine and sufficiently serious threat to the environment would thus fall squarely within the concept of ordre public', para. 109 (Advocate General Jacobs).

112 Article 27(2).

113 *Plant Genetic Systems,* above n. 108, para. 5.

Although considered together with ordre public by the EPO, the morality proviso is given a distinctive meaning:

> The concept of morality is related to the belief that some behaviour is right and acceptable whereas other behaviour is wrong, this belief being founded on the totality of the accepted norms which are deeply rooted in a particular culture . . . the culture in question is the culture inherent in European society and civilisation.[114]

The very idea of an all-embracing concept of morality, determinable by patent officers, seems somehow old-fashioned, even authoritarian. However, the openness of the morality exception means that a full range of issues can potentially be raised here.

Some inventions are deemed by the Biotechnology Directive to be contrary to ordre public/morality:

> (a) processes for cloning human beings;
> (b) processes for modifying the germ line genetic identity of human beings;
> (c) uses of human embryos for industrial or commercial purposes;
> (d) processes for modifying the genetic identity of animals which are likely to cause them suffering without any substantial medical benefit to man or animal, and also animals resulting from such processes.[115]

Whilst it is not bound by the Directive, the EPO aligned its practice with this part of the Biodiversity Directive by introducing a rule of interpretation incorporating these examples. Only the final paragraph (d) potentially applies to GMOs, and that only to animals modified and farmed for medical purposes, for example to express pharmaceutical products in their milk. Most farmed animals and plants will be assessed on a free-standing basis, under the basic ordre public/morality exception. For current purposes, however, these explicit exceptions from patentability demonstrate at least that the morality provision is intended to be meaningful – there is no 'right' to a patent regardless of the consequences.

The EPO is generally perceived and perceives itself as an expert-technical body, outside the realm of politics. The patent system is, however, unusually open to external challenge. Under the EPC, not only can a disappointed patent applicant appeal a decision (as in the first round of the *Oncomouse (Europe)* saga), but 'any person' can initiate opposition proceedings against the grant of a patent (as in the final round). Whilst we might normally expect such

---

114 *Plant Genetic Systems*, above n. 108, para. 6, cited *Harvard*, above n. 100, para. 10.2.

115 Directive 1998/44, above n. 6, Article 6(2).

proceedings to be initiated by a commercial competitor, this provides access to outsiders including environmental interest groups such as Greenpeace, animal welfare/rights groups such as the British Union for the Abolition of Vivisection (BUAV) and Green Members of the European Parliament.[116] It is striking to note that such a free-ranging opportunity for legal challenge is not found anywhere else in the regulatory process for GMOs. Whilst there are opportunities to comment within the authorisation process, for example, standing rules make legal challenge at EU level virtually impossible for a public interest group such as Greenpeace.[117] The range of legitimate inquiry once the outsiders are before the EPO also compares well in principle with the 'risk regulation', primarily because of this far-reaching ordre public/morality exception. The extent to which a decision can step beyond the purely technical is, however, determined by the precise application of this exception. We might recall how the potential breadth of the 'other legitimate factors' formula in the Food and Feed Regulation is restricted by the political and legal context of decision making.[118]

Moving on to specific applications of the ordre public/morality exception, the final instalment of the tortuous *Oncomouse (Europe)* case was a decision by the Technical Board of Appeal (TBA) in 2004. This was the second time the oncomouse had reached the TBA: earlier, it had heard an appeal against the refusal of a patent by the Examining Division,[119] remitting the application back to the Examining Division, which on this second look had granted the application. The final TBA decision in 2004 discussed the morality exception at some length.[120] It applied the explicit exception of paragraph (d) above first. This requires consideration of likelihood of suffering and likelihood of substantial medical benefit. In this case, the TBA found that suffering is not

[116] Greenpeace in *Plant Genetic Systems*, above n. 108, BUAV in *Harvard,* above n. 100, and a group of Members of the European Parliament in *Relaxin*, above n. 27.

[117] See the discussion in Maria Lee, *EU Environmental Law: Challenges, Change and Decision-Making* (Hart Publishing, 2005), Chapter 5. Regulation 1367/2006/EC on the application of the provisions of the Aarhus Convention on Access to Information, Public Participation in Decision-making and Access to Justice in Environmental Matters to Community institutions and bodies [2006] OJ L 264/13 provides opportunities for environmental interest groups to require the Commission to reconsider its decision, with review by the Court. It does not necessarily, however, provide for judicial review of the original decision.

[118] Regulation 1829/2003 on Genetically Modified Food and Feed [2003] OJ L 268/1. See the discussion in Chapter 3.

[119] T-19/90 *Harvard/Oncomouse* [1990] EPOR 501.

[120] See the discussion in David Thomas and Georgina A. Richards, 'The Importance of the Morality Exception Under the European Patent Convention: The Oncomouse Case Continues' (2004) 26 *European Intellectual Property Review* 97. Also at issue were questions of an 'inventive step' and disclosure.

only likely but inevitable. Equally, however, substantial medical benefit in respect of mice was also established.[121]

Of more direct relevance for current purposes, the TBA moved on to consider what it referred to as the 'real' Article 53(a) objection, the more general ordre public/morality provision. It applies the approach taken in its first oncomouse decision:

> The decision as to whether or not Art 53(a) EPC is a bar to patenting the present invention would seem to depend mainly on a careful weighing up of the suffering of animals and possible risks to the environment on the one hand, and the invention's usefulness to mankind on the other.[122]

This balancing of harm and usefulness is in principle 'sufficiently flexible to allow for the current . . . views as to social order, environmental risk and accepted standards of behaviour in European culture'.[123] This could be an important opportunity to engage with whether the *benefits* of the GMO, or the social *need* for any particular manifestation of agricultural biotechnology merit the costs or uncertainties. The triviality or uncertainty of the benefit is not, it will be recalled, an explicit element of 'risk regulation' in the EU. Whilst these principles appear to be broad, however, their application is rather restrictive. So, for example, it is apparently necessary to establish an alleged risk to the environment by the virtually impossible standard of 'conclusive evidence' that it is 'likely to seriously prejudice the environment', a standard unthinkable in 'ordinary' environmental regulation.[124] In the case of the oncomouse, the environmental risk was regarded as only 'minimally more than hypothetical', taking into account both the existence of a regulatory framework to reduce the possibility of escape and the conclusion that it is 'questionable' in any event whether escaped mice would 'cause any damage, let alone any lasting damage' to the environment.[125] The EPO's willingness to defer to subsequent regulators, together with its very high evidential demands, make for an extremely restrictive approach to the ordre public/morality exception.

Article 53(a) is not, however, limited to this balancing of proven harm against benefits. EPO Guidelines consider it a 'fair test' of the morality provision 'to consider whether it is probable that the public in general would regard

121 But not in respect of all rodents ('squirrels, beavers, porcupines and every other rodent'), para. 12.2.1. This shows the importance of the drafting of a patent claim.

122 *Harvard*, above n. 100, para. 10.5.

123 *Harvard*, above n. 100, para. 10.5.

124 *Harvard*, above n. 100, para. 13.2.8, citing *Plant Genetic Systems*, above n. 108, para. 18.7.

125 *Harvard*, above n. 100, para. 13.2.9.

the invention as so abhorrent that the grant of patent rights would be inconceivable'.[126] This demanding standard applies to 'inventions that would universally be regarded as outrageous',[127] and only in 'rare and extreme cases'.[128]

This takes us back to the broader definition of morality, mentioned above, as 'the totality of the accepted norms which are deeply rooted in a particular culture . . . the culture in question is the culture inherent in European society and civilisation'.[129] Just as there is unlikely ever to be a single 'public opinion' to follow (Chapter 2), however, unified *European* moral judgments are likely only in the most extreme of cases, if ever. The morality exception in the Biotechnology Directive is directed at the Member States rather than the supranational level.[130] The ECJ in *Netherlands* stated that 'national legislative, administrative and court authorities are better placed to understand' the different 'social and cultural context' in different countries and localities.[131] Advocate General Jacobs, in his Opinion in the same case, observed that 'some ethical issues may be more appropriately evaluated in the context of the culture of a particular Member State and others are susceptible to a common standard', and seemed to prefer a 'degree of harmonisation'.[132] The identification of a single European moral judgment would though be highly attractive to the EU institutions. The Biotechnology Directive states rather baldly that 'The Commission's European Group on Ethics in Science and New Technologies evaluates all ethical aspects of biotechnology',[133] suggesting at least that this expert group will make a contribution to the ordre public/morality assessment. This potentially moves towards the creation of a single European solution to a moral dilemma *defined* as a European problem. The scale of the morality judgment (European, national or local) in principle remains open, but there is a real impetus towards 'Europeanisation'. There is potential further to squeeze out diversity here, as well as to reinforce the tendency to render increasing amounts of ordinary political discourse a technical matter for resolution by experts.

---

126 *Guidelines for Examination in the European Patent Office* (European Patent Office, 2005).

127 *Relaxin*, above n. 27, para. 6.2.1.

128 Guidelines, above n. 126.

129 *Plant Genetic Systems*, above n. 108. Cited in *Harvard*, above n. 100, para. 10.2.

130 '*ordre public* and morality correspond in particular to ethical or moral principles recognised in a Member State, respect for which is particularly important in the field of biotechnology in view of the potential scope of inventions in this field and their inherent relationship to living matter' (Recital 39).

131 *Netherlands*, above n. 17, para. 38.

132 *Netherlands*, above n. 17, para. 102.

133 Directive 1998/44, above n. 6, Article 7.

The location of moral judgments remains unclear not only in scale. Arguments advanced under Article 53(a) need to be substantiated. Evidence on moral acceptability might, in addition to looking to experts, be sought in national or EU laws and regulations. The fact that something is unlawful does not in itself, however, mean that it cannot be patented under the ordre public/morality exception,[134] and vice versa.[135] This leaves some scope for independent judgment within the patent system. But neither the EPO nor the ECJ is clear on the degree to which morality can be elided with public opinion. In its *Oncomouse (Europe)* decision, the TBA found insufficient evidence that the public found the use of genetically manipulated mice in medical research morally unacceptable. The EPO treats opinion poll evidence with some suspicion, identifying a number of queries about the information provided, including the type and the number of questions posed within one poll, the size and representativeness of the polling group, the interpretation of results,[136] and the methodology of the polls.[137] Some caution is appropriate – opinion poll evidence can be shallow and might easily be manipulated. But, notwithstanding the difficulties, it is hard to see what is at issue in the morality exception if some effort to gauge public views is not involved. The setting of standards for acceptable evidence on public opinion is hardly inconceivable, and the queries raised by the EPO are capable of response. Decision making on matters of high politics is not a case of counting heads, but of identifying concerns and engaging with them. As has been observed elsewhere in this book, whilst scientific evidence is preferred for its apparent objectivity and neutrality, it is impossible to escape uncertainty and impossible to escape value judgments, simply by recourse to one form of expertise rather than public views.[138] To conclude on

---

[134] EPC Article 53(a); Biotechnology Directive Article 6(1). See the Opinion of Jacobs AG in *Netherlands*, above n. 17, especially para. 106.

[135] *Plant Genetic Systems*, above n. 108. *Harvard*, above n. 100, however, does seem to use the legislative context as *evidence* that both animal welfare and the use of animals in research are 'established in the European culture', para. 1.2.18.

[136] *Harvard*, above n. 100, para. 13.2.4, citing *Plant Genetic Systems*, above n. 108. See also *Relaxin*, above n. 27, in which the opponents had requested the EPO to carry out a referendum on the subject, para. 6.5.

[137] 'for example, whether [the polls] were conducted by trained professional pollsters or by casual staff . . . whether the respondents were stopped on street corners and answered questions in a hurry or were invited into comfortable premises and given time to think; whether they were volunteers or were paid for participating; what other questions they were asked as well as those specifically relied on in these proceedings . . . whether the questions they were asked were "open" . . . or "closed" . . .; and how the results were analysed', *Harvard*, above n. 100, para. 203.

[138] See also Amanda Warren-Jones, 'Identifying European Moral Consensus: Why are the Patent Courts Reticent to Accept Empirical Evidence in Resolving Biotechnological Cases?' (2006) 28 *European Intellectual Property Review* 26.

Oncomouse (Europe), the TBA rejected the arguments that the morality exception should be applied: the 'unease' that it accepted existed could not 'be elevated to the status of moral disapproval in European culture of the use of animals for medical research, let alone moral disapproval of the use of mice in cancer research'.[139]

As in other areas of agricultural biotechnology, the moral dilemma is more pronounced, and perhaps in some cases more compelling, with respect to animals. It is not surprising that the most thorough discussion should have arisen in this context. It is virtually uncontentious that animal suffering should not be imposed without reason, although what counts as a reason, and whether there are any acceptable reasons, is a matter of disagreement. The oncomouse patent survived assessment by reference to external standards of ordre public and morality in the EPO, but only after debate. It is the debate in itself that is striking in the biotechnology arena.

More difficult, and extending to plants, would be the discussion of the sorts of biocentric or religious perspectives outlined in Chapter 2. It has been argued before the US Congress that patents can reflect 'a human arrogance towards other living creatures' that denies 'the inherent sanctity of every unique being and the . . . ecological and spiritual inter-connectedness of all life'.[140] Before the EPO, Greenpeace challenged the morality of 'the dominion that was sought to be exercised by man over the natural world by the use of plant genetic engineering techniques'. The EPO characterised these concerns as 'understandable . . . because the power of science for good and evil has always troubled man's mind'. According to the EPO, the question is whether the claim relates to a 'misuse or a destructive use' of plant biotechnology, and in this case it finds that the use of biotechnology cannot be considered to be 'wrong as such'.[141] It is hardly surprising that the EPO does not really engage with the non-negotiability of the biocentric position, and it does at least acknowledge the existence of the concern albeit in rather a condescending manner. Although the existence of the Biotechnology Directive suggests that a blanket ordre public/morality exception for biotechnological inventions or living organisms generally is not legally appropriate, it is apparently legitimate to think about what is involved for the intrinsic value of living organisms by patenting any particular invention.[142]

Proper consideration of the moral impact of the distributional concerns discussed earlier in this chapter and in Chapter 2 is also quite striking by its

139 *Harvard*, above n. 100, para. 13.2.21.

140 Humane Society of the United States, submissions to Congress, cited in Kevles, above n. 97, pp. 49–50.

141 *Plant Genetic Systems*, above n. 108, paras 17.1 and 17.3.

142 See the discussion in Chapter 2.

absence in these cases. Patenting and associated monopoly rights are most controversial in respect of life's necessities. So the pharmaceutical industry has faced enormous criticism for standing on its rights in the face of public health crises in the developing world, and control of food agriculture obviously also falls into this category. Perhaps something more imaginative than the simple yes/no response of the ordre public/morality exception would be helpful here – 'normal' regulation rarely deems non-conditional authorisation to be the only alternative to a denial of authorisation. Some combination of the measures being considered in respect of disclosure and equitable sharing of benefits might be most appropriate.

There is huge potential in the patent system for a real airing of the broad range of concerns about agricultural biotechnology. That potential is resisted even more explicitly than in Europe in jurisdictions without the ordre public/morality exception. In *Diamond v Chakrabarty*, the Supreme Court was presented with 'a gruesome parade of horribles', including 'a serious threat to the human race', pollution and disease, loss of genetic diversity and the depreciation of 'the value of human life'.[143] The Court argued that it was 'without competence to entertain these horribles, either to brush them aside as fantasies generated by fear of the unknown, or to act on them'.[144] These were held to be matters of 'high policy' for political resolution. Except, of course, that by its silence the Supreme Court necessarily provided a resolution in practice. Moreover, the Canadian Supreme Court majority expressed similar concerns:

> this Court does not possess the institutional competence to deal with issues of this complexity, which presumably will require Parliament to engage in public debate, a balancing of competing societal interests and intricate legislative drafting.[145]

However, Canada reached the opposite conclusion from the US, deciding that higher life forms are not patentable. The Supreme Court minority in Canada would have held that the narrow legal issue 'does not provide a proper platform on which to engage in a debate over animal rights, or religion, or the arrogance of the human race'. This is of course fair, but it is clear even from the dissent itself, which discusses issues such as the commercial and social benefits of biotechnology for Canada, that values are necessarily involved in any event.[146]

143 Above n. 97, p. 2211.
144 Above n. 97, p. 2212.
145 *Harvard College*, above n. 76, para. 183.
146 *Harvard College*, above n. 76, *per* Binnie J. He is aware of the dilemma: 'This is not to suggest that because something is beneficial it is necessarily patentable . . . such value judgments have been excluded from the administration of the *Patent Act*', para. 18.

It is often argued that objections to the scope of patent protection are actually not about patent law but about the undesirability of the research in the first place – and that the appropriate place for regulation is not in patent law but in 'regulation'. Philip Grubb makes the classic case:

> If it is felt that it is immoral to make transgenic animals or human embryonic stem cell lines, then surely it would be more logical to concentrate on campaigning for legislation to stop or restrict such experiments, rather than trying to lock the door after the horse has bolted by stopping patents for work which has already been done.[147]

Turning to other forms of control makes sense, and the grant or denial of a patent certainly should not be the full extent of regulation of new technologies. The grant of a patent does not allow the use of a product of biotechnology; it merely grants a monopoly on exploitation. So the denial of a patent, in the absence of other regulation, would mean that anyone could exploit the invention, not that no one would exploit. The recitals to the Biotechnology Directive remind us of this, and put the 'regulatory' framework centre stage, especially with regard to 'the requirements of public health, safety, environmental protection, animal welfare, the preservation of genetic diversity and compliance with certain ethical standards'.[148] The attention given to patents by those opposed to biotechnology does mean that the intellectual property field is being asked to shoulder rather a large part of the burden of reconciling the risks and benefits of new technology. That does not mean, though, that patents have no role at all. To argue that the justifiability of an invention is a matter only for 'other' regulators fails to acknowledge the claimed benefits of patents in incentivising innovation.

Grubb goes on from the quotation above to outline some of the specific ethical dilemmas associated with human gene lines: informing people of a genetic risk of disease when they can take no protective action, the availability of genetic information to employers and insurers, the compulsory collection of DNA from convicted criminals. 'These are the areas in which the real ethical dilemmas are to be found, and arguments about whether or not one should patent a gene are completely irrelevant to these real issues.'[149] But, to take the most extreme example, it is far from clear that the patenting of an invention arising out of the compulsory collection of DNA is a neutral act. The appropriate nature of ownership in such information, and the propriety of awarding the public benefit of a monopoly right, are at the very least subjects

147 Grubb, above n. 8, p. 285.
148 Directive 1998/44, above n. 6, Recital 14.
149 Grubb, above n. 8, p. 271.

on which reasonable people might disagree, and the legitimate topic of ethical debate. This returns the ethical dilemma very firmly to the patent sphere, albeit without removing it from the other areas. Grubb argues that objectors to patents in biotechnology simply fail to understand the patent system. Alternatively though, they might dispute the neutrality of the system, and its implicit reward for inventions the legitimacy of which could at least be the subject of debate.

Patent professionals may well not be our first choice of morality examiners. But, whilst they are often keen to disclaim competence to assess moral questions, the greater danger, rather than overreaching, is perhaps that patenting is misconceived as a purely technical or legal issue. The patent system necessarily provides and applies an ethical framework, but, in the shared technical assumptions of the patent world, that can go unacknowledged. The moral perspective of patent law is usually directed to the promotion of economic and commercial interests which are deemed to be in the public interest. But the patent system has always been used to reflect broader moral concerns: declining to patent contraceptives (until contraception was perceived to be in the public interest) or gambling devices are important examples,[150] as are the legislative examples of inventions contrary to ordre public/morality. On this subject, we might also note concern in the EU that the patenting of tests for the presence of GMOs will mean that only the industry will be able to check whether there is GM material in products. Legislative intervention means that the use of 'detection methods' and 'reference materials' provided under the process for authorisation of GMOs 'shall not be restricted by the exercise of intellectual property rights or otherwise'. Nor, importantly, will such detection and identification methods be deemed to be confidential.[151] This limit on the scope of patent protection places testing in the public domain, making the direct regulation of GMOs realistic.

CONCLUSIONS

The context within which the authorisation of GMOs in the EU is assessed includes rules on intellectual property. These rules are part of the regulatory settlement, part of the complex balance between encouraging economic development through biotechnology and regulating for the public interest. There is a strong belief in the EU institutions in the positive link between a strong

150 Angus Wells, 'Patenting New Life Forms: An Ecological Perspective' (1994) 16 *European Intellectual Property Review* 111.

151 Regulation 1829/2003 above n. 118, Article 30(5) and (3)(f).

patent system and a healthy economy, and patenting in particular emphasises enthusiasm for the positives of biotechnology. As a part of the regulatory framework, patent law plays its role in shaping the relationships between those affected by agricultural biotechnology, but it focuses on encouraging and rewarding innovation rather than on control, assuming that innovation is in the public interest. The main purpose of this chapter has been to demonstrate the ways in which patent law contributes to the regulatory framework for agricultural biotechnology, distributing goods and bads at global and neighbourhood levels, and making implicit or explicit value judgments about technical progress.

Importantly, patent law also provides a space in which to discuss the appropriateness of certain technological developments. It is edifying to observe the expansion of the scope of the debate on the application of patent rules to biotechnology. The debate is no longer focused solely on the narrowly technical questions, for example as to whether the subject matter of the patent claim constitutes an 'invention' in legal terms or just a (non-patentable) 'discovery', but has been extended to much more expansive questions as to the role and impact of patenting, and from a narrow field of specialists to a broader group of experts and publics.[152] We have observed a similar widening of relevance in the more general debate on agricultural biotechnology, from simply a scientific debate about risk to an explicitly ethical and socioeconomic debate. In both cases, the legal and political contexts of decision making mean that actual engagement with these broader issues is difficult, in spite of acceptance of their legitimacy in principle. The morality exception in patent law is a far more direct approach than the tangential references to ethical considerations and 'other legitimate factors' in the authorisation process. But, as in other spheres of regulation, outsider perspectives still struggle to be heard. In the circumstances of competition between the major global economies, Drahos characterises the morality exception as a 'gently floating irrelevance'.[153]

The desperate search for a forum for 'moral' debate has pushed social discussion into the patent arena. Some forum for the consideration of social and ethical issues associated with new technology is a necessary part of its regulation. I would not wish to disagree with those who argue that patent lawyers are not best placed to pronounce on morality. There is, however, no obviously preferable alternative available. We could hive social and ethical issues off into a separate and self-contained forum, but that contributes to the illusion of objectivity in traditional fields of law (intellectual property, tort,

152 See Geertrui van Overwalle, 'Reshaping Bio-patents: Measures to Restore Trust in the Patent System' in Somsen, above n. 50, for a discussion of this shift of approach.

153 Above n. 23.

risk regulation). This debate should be kept within the mainstream. That means that a proper consideration of the incentivisation of development through intellectual property rights is important for anyone interested in the 'regulation' of GMOs.

6. The global context of international trade

INTRODUCTION

The most striking element of the World Trade Organization (WTO) Director General's Yale speech on 'trade and sustainable development' is not his light treatment of a complex issue, or even his complacency about environmental impacts ('The GATTs have already been relatively greened, and if we accomplish the Doha Round, we would green them some more!'). Most striking is Pascal Lamy's very easy acceptance that the 'trade and sustainable development' debate is about 'values that could cross national borders'.[1] This acceptance that trade is not just about trade, and is not simply a technical exercise that can be isolated from politics or the pursuit of other social goods, has been rather hard won. But quite what to do with this knowledge is still difficult. Whilst the WTO does conscientiously provide space for members to explain measures falling foul of trade rules on the basis of non-economic values, some of those explanations are scarcely heard in an overwhelmingly trade-oriented, liberal economic institution.

A workable international consensus on the appropriate role and regulation of GMOs seems very far away. The EU's caution contrasts sharply with the rapid adoption of the technology in the US and elsewhere. The continued and profound disagreement between these two major economic powers creates real (and realised) potential for international conflict. GMOs are an enormous challenge for the WTO. Respecting the response of democratic systems to citizen concerns, whilst simultaneously maintaining the integrity of international trade rules, is a delicate task, and disagreement over GMOs has massive potential to spill over into broader debates about the legitimacy of the WTO. The EU also faces serious risks as GMOs enter the international arena, with the WTO context likely to amplify disagreement at EU level. The EU could show itself to be incapable of delivering, that is, incapable of regulating in its

[1] Director General Pascal Lamy, *The WTO and its Agenda for Sustainable Development* (Yale University, 24 October 2007) available at http://www.wto.org/english/news_e/sppl_e/sppl79_e.htm (accessed December 2007).

jurisdiction. Or it could show itself to be incapable of protecting its own citizens from risks, including the risks of being part of a global trading system. The influence of the WTO has been apparent in the development of the EU regime on GMOs[2] and indeed the EU context has also had its effect in the US,[3] the EU's main antagonist in the WTO. The EU has simultaneously strongly asserted its right to autonomous decision making in this area.

Three elements of the EU moratorium on authorisation of GMOs have been challenged before the WTO by the US, Canada and Argentina: the general moratorium, the failure to make decisions in respect of a number of specified products and certain Member State safeguard measures. In its 2006 decision, the *EC–Biotech* Panel found that in all three challenged categories the EC had violated the Sanitary and Phytosanitary (SPS) Agreement.[4] The EU's behaviour during the moratorium, unsubtle and unplanned, with no explanation and no real opportunity for those affected to challenge (non-)decisions, would be difficult to defend in any circumstances. Nevertheless, the reasoning of the Panel on this extraordinarily sensitive and difficult issue was in places unnecessarily reductive. The EU did not appeal, but this is far from the end of the saga. *EC–Biotech* did not address the new EU regulatory system for GMOs, discussed in detail in Chapters 3 and 4 above. Assuming that this new system does eventually function effectively, its onerous regulatory requirements in respect of authorisation, labelling and traceability are likely to raise novel and significant questions about WTO law.[5]

Any of the three most far-reaching WTO Agreements (the General Agreement on Tariffs and Trade (GATT), the SPS Agreement and the Agreement on Technical Barriers to Trade (the TBT Agreement)) could be used to assess EU regulation of GMOs. Simply put, the GATT imposes a non-discrimination framework that requires that imports from WTO members be treated no less favourably than imports from other members (the 'most

---

2 See Gráinne de Búrca and Joanne Scott, 'The Impact of the WTO on EU Decision-making' in Gráinne de Búrca and Joanne Scott (eds), *The EU and the WTO: Legal and Constitutional Issues* (Hart Publishing, 2001), on the potentially chilling impact of the WTO on the development of EU law.

3 Particularly by alerting consumers and environmental groups to the controversy; see Alasdair R. Young (2001), *Trading Up or Trading Blows? US Politics and Transatlantic Trade in Genetically Modified Food* EUI Working Papers, RSC 2001/30.

4 *EC – Measures Affecting the Approval and Marketing of Biotech Products,* (DS291, 292, 293, 29 September 2006).

5 There is a large literature on the subject. See particularly Joanne Scott, 'European Regulation of GMOs and the WTO' (2003) 9 *Columbia Journal of European Law* 213; Robert Howse and Petros C. Mavroidis, 'Europe's Evolving Regulatory Strategy for GMOs: The Issue of Consistency with WTO Law: of Kine and Brine' (2000) 24 *Fordham International Law Journal* 317.

favoured nation' principle, Article I) and that imported products be treated no less favourably than domestic products (the 'national treatment' principle, Article III). The SPS Agreement disciplines the use of sanitary and phytosanitary (basically, human, animal and plant life and health) measures. The TBT Agreement applies to 'technical regulations', which includes the regulation of 'product characteristics' and mandatory labelling requirements.[6] The WTO is a youthful system, and predicting precisely which parts of which agreement will be applied to which elements of the regulatory framework for GMOs is not easy. The GATT and the TBT Agreement impose obligations that can both apply to the same measure, and complying with one does not preclude application of the other.[7] These two agreements are highly likely to be applied to parts of the new regulatory regime for GMOs, but the main agreement, for reasons explored further below, is likely to be the SPS Agreement.[8] The SPS Agreement applies independently of any breach of the GATT, although conformity with the SPS Agreement implies conformity with the GATT. If the SPS Agreement applies, there is no residual space for the TBT Agreement (although the same measure can constitute, independently, an SPS and a TBT measure, below p. 202).

Rather than attempting a detailed analysis of the possible application of WTO law to the regulation of GMOs, this chapter examines some of the key quandaries. Perhaps most obviously, the very understanding of agricultural biotechnology as a trade issue, rather than a social, distributional or environmental issue, is problematic. Whilst international trade can bring all sorts of benefits and advantages for the world, most simply economic growth, there are important questions to be asked about how other values stand up to trade's predominantly economic values. This chapter begins by examining, in the next two sections, the explicit provision for inclusion of broad social values in the WTO system, as well as the WTO's potential to enhance democratic account-

---

6 A technical regulation is a 'Document which lays down product characteristics or their related processes and production methods, including the applicable administrative provisions, with which compliance is mandatory. It may also include or deal exclusively with terminology, symbols, packaging, marking or labelling requirements as they apply to a product, process or production method', TBT Agreement, Annex I. The Appellate Body considered this in *EC – Measures Affecting Asbestos and Asbestos-Containing Products* (WT/DS135/AB/R, 12 March 2001), para. 66 and onwards. Technical standards, which are not mandatory, are also covered by the TBT Agreement.

7 *Asbestos*, above n. 6. The breadth of scope of the TBT and SPS Agreements does raise questions about precisely what role is left for the GATT.

8 See especially Joanne Scott, *The WTO Agreement on Sanitary and Phytosanitary Measures: A Commentary* (Oxford University Press, 2007), for an interesting account of the agreement.

ability. These two sections concentrate on the GATT and the TBT Agreement. The SPS Agreement also provides space for the pursuit of social values, primarily of safety, but the constraints imposed by the science-based disciplines of the SPS Agreement are considerable, and are discussed in detail at pp. 211–22. The *EC–Biotech* decision is an important reference point in this discussion of the role of science and risk assessment in the WTO. Another significant feature of this decision are the restrictions it imposes on the ability of regulators to take time under the SPS Agreement to understand and address public responses to a technology, and this element of time and politics is also discussed below. Whilst the moratorium, as already mentioned, is an exceptional rejection of legal expectations, the Panel's failure to grasp properly the EU's fragile authority at this time is worrying. The role of other international agreements and standards in the WTO system is also significant in the future of GMO regulation, and that will be explored before turning finally to the possible application of the GATT and the TBT Agreement to the EU's new regime.

THE WTO AND SOCIAL VALUES

A recurring question in this book has been the ability of the regulatory process to hear the social, ethical, economic and political dilemmas of agricultural biotechnology. A whole range of public goods and social values do find a place in the WTO regime, and this would be easy to miss from listening to more passionate critics of the WTO. The role of these social values is, however, constrained. Most fundamentally, they always have to compete with the assumed pre-eminence of basic free-trade rationales. This places the alternative social values in a subsidiary position from which their purpose always has to be argued, and in a context that places serious disciplines on their use.[9] Whether the full breadth of possible social concerns about GMOs (see especially Chapter 2) will resonate at WTO level is open to question.

The preambles to all three of the WTO agreements discussed here emphasise non-trade social objectives. The preamble to the GATT, which provides 'colour, texture and shading' to the operative provisions,[10] clarifies that trade is not an end in itself:

---

9 See Robert Howse, 'From Politics to Technocracy – and Back Again: The Fate of the Multilateral Trading Regime' (2002) 96 *American Journal of International Law* 94, for a provocative discussion of the evolution of the struggle (or deal) between freer trade and (broadly) the welfare state.

10 *United States – Import Prohibition of Certain Shrimp and Shrimp Products* (WT/DS58/AB/R, 12 October 1998), para. 153.

> *Recognizing* that their relations in the field of trade and economic endeavour should be conducted with a view to raising standards of living, ensuring full employment and a large and steadily growing volume of real income and effective demand, and expanding the production of and trade in goods and services, while allowing for the optimal use of the world's resources in accordance with the objective of sustainable development, seeking both to protect and preserve the environment and to enhance the means for doing so in a manner consistent with their respective needs and concerns at different levels of economic development.

The most compelling place within the GATT for discussion of social values is in Article XX, which provides exceptions to the general GATT rules. Exceptions are provided, inter alia, for measures 'relating to the conservation of exhaustible natural resources if such measures are made effective in conjunction with restrictions on domestic production or consumption' (Article XX(g)), measures 'necessary to protect human, animal or plant life or health' (Article XX(b)) and measures 'necessary to protect public morals' (Article XX(a)).

The Appellate Body's application of Article XX(g) in the *Shrimp/Turtle* dispute is well known. The US had introduced restrictions on the import of shrimp, in an effort to reduce the death of turtles by shrimp fishing. The measures were complex, but essentially only shrimp that the US was satisfied had been fished using 'turtle excluder devices' could be sold in the US. The Appellate Body accepted the legitimacy of the US's objectives. The turtles did constitute 'exhaustible natural resources' for purposes of Article XX(g), and, without determining whether there is an 'implied jurisdictional limitation' to Article XX(g), the Appellate Body held that in this case 'there is a sufficient nexus between the migratory and endangered marine populations involved and the United States', at least in part because the species in question did all occur in US waters.[11] Although drafted at a time when environmental protection had almost no policy resonance, Article XX(g) is read 'in the light of contemporary concerns of the community of nations about the protection and conservation of the environment'.[12] Conservation objectives are clearly legitimate within the GATT framework, and Article XX has gone beyond charismatic endangered species like these turtles. So paragraph (g) has been applied to non-endangered dolphins[13] and to clean air.[14] The inclusion of routine envi-

---

11 *Shrimp/Turtle*, above n. 10, para. 133.

12 *Shrimp/Turtle*, above n. 10, para. 129.

13 *United States Restrictions on Imports of Tuna* (1992) 30 ILM 1598 (*Tuna/Dolphin* I); *United States Restrictions on Imports of Tuna* (1994) 33 ILM 839 (*Tuna/Dolphin* II).

14 *United States – Standards for Reformulated and Conventional Gasoline* (WT/DS2/AB/R, 29 April 1996).

ronmental objectives in Article XX(g) is relatively secure, backed up by Article XX(b), which has been interpreted to embrace 'environmental protection', when that is 'shorthand' for the 'animal or plant life or health' explicitly covered in that paragraph.[15] This was confirmed in an EC challenge to Brazil's ban on the import of retreaded tyres. Brazil argued that such imports (with a shorter lifespan than new tyres) increase the generation of waste tyres, and hence associated health and environmental problems, including water pollution, the provision of breeding grounds for mosquitoes (and hence mosquito-borne diseases such as malaria, yellow fever and dengue fever) and fires that are difficult to control and produce hazardous emissions.

The idea of a 'public morals' exception is familiar from discussion of patents in the previous chapter. The patents exception was really only thrust into the limelight with the development of biotechnology, and it is possible that the public morals exception of the GATT will similarly be called on as biotechnology returns to the trade arena. The WTO dispute settlement bodies are, however, closed institutions relative to the European Patent Convention (EPC). Under the EPC, it will be recalled, 'any person' can bring opposition proceedings. In the WTO, the member (that is, the state or the EU) has the prerogative of action. States are perhaps less likely to rock the boat with wide-ranging and unpredictable debates about public morals than single-issue public interest groups desperate for a forum in which to express their concerns. And as discussed below (pp. 209–10), WTO panels exercise their discretion to hear outsider perspectives sparingly. Whilst interest group coalitions do seek to provide help to the dispute settlement bodies via amicus briefs, their desire to be heard in a relatively hostile environment presumably provides strong incentives for the framing of arguments in the most conventional and least frightening way possible.

There is only one example of the Appellate Body considering a 'morality' exception under the WTO Agreements. The *Gambling* dispute arose out of US restrictions on internet betting, and was brought by Antigua under the General Agreement on Trade in Services (GATS).[16] Article XIV(a) of GATS contains a morality clause similar to that found in the GATT, and subject to a similar 'chapeau':

> Subject to the requirement that such measures are not applied in a manner which would constitute a means of arbitrary or unjustifiable discrimination between

---

[15] *Brazil – Measures Affecting Imports of Retreaded Tyres* (WT/DS332/R, 12 June 2007), para. 7.46 (Panel Report). Note also the interpretation of the SPS Agreement to include environmental issues, below, pp. 234–5.

[16] *United States – Measures Affecting the Cross-Border Supply of Gambling and Betting Services* (DS285/AB/R, 7 April 2005).

> countries where like conditions prevail, or a disguised restriction on trade in services, nothing in this Agreement shall be construed to prevent the adoption or enforcement by any Member of measures:
> (a) necessary to protect public morals or to maintain public order.

The Appellate Body held that this should be interpreted in the same way as the public morals provision in the GATT.[17] It upheld the Panel's approach to morality, which had been to interpret 'public morals' as '[denoting] standards of right and wrong conduct maintained by or on behalf of a community or nation'.[18] Concerns about money laundering, organised crime, fraud, underage gambling and pathological gambling constitute 'very important societal interests'.[19] The *Gambling* decision is of limited assistance in working out the relevance of the ethical context surrounding GMOs. The embrace of the US measures by the public morals exception was very briefly dealt with, to the extent that both the Appellate Body and the Panel came close simply to assuming that the objective was public morals protection. This broad approach to the reach of 'public morals', together with the Panel's observation that the content of these standards 'can vary in time and space, depending upon a range of factors, including prevailing social, cultural, ethical and religious values'[20] helps to avoid any second-guessing of domestic decisions on the moral considerability of problems within the jurisdiction. The Panels, and especially the Appellate Body, are conscious of how unappealing such second-guessing would be. In the context of Article XX(b), a Panel is not 'required to examine the desirability of the declared policy goal as such'.[21] Although Article XX(b) is a more tightly drawn exception (the protection of human, animal or plant life and health is explicitly deemed acceptable by the GATT itself) a cautious approach would be wise even in respect of the more open-ended public morals provision. This is not to argue that simple assertion of public morals should mean that there is no breach of the GATT, but simply that the discussion should turn relatively easily to Article XX(a). The pursuit of even legitimate public morals objectives is then controllable by the necessity proviso and the chapeau.

---

[17] There is though an obvious difference, on which nothing turns for current purposes. The term 'public order' does not appear in the GATT. A footnote to paragraph (a) provides that 'the public order exception may be invoked only where a genuine and sufficiently serious threat is posed to one of the fundamental interests of society'; see *Gambling*, above n. 16, para. 296.

[18] *Gambling*, above n. 16, para. 296.

[19] *Gambling*, above n. 16, para. 310.

[20] *Gambling*, Panel Report (WT/DS285/R, 10 November, 2007), para. 6.461.

[21] *Brazil–Tyres*, above n. 15, para. 7.97, Panel Report.

The reach of the morality exception is not however likely to be completely without bounds. A public morals exception seems reasonably comfortably to apply to religious restrictions, for example on the consumption of alcohol or certain food, or even control of genetic modification that produces animal (especially pig or cow) genes where they are not usually expected. Gambling, pornography, and narcotics are also relatively often placed in the bracket of moral concerns. Whilst a particular approach to, for example, alcohol, cannot be *applied* across cultures or religions, it can be easily recognised and understood. And it is likely that implicit consensus or explicit international agreement will help: recall the Appellate Body's interpretation of Article XX(g) 'in the light of contemporary concerns of the community of nations'.[22] Turning specifically to GMOs, public morals might extend even to concerns about animal welfare, interfering with nature, the ethics of taking poorly understood risks,[23] the distributional impacts of regulation and a consumer 'right to know'. But domestic approaches to these types of issues may have little moral significance outside their own cultural space, and will be far more difficult to 'fit' into the WTO context. It is unlikely that the Appellate Body will allow Article XX(a) to turn into a kind of 'catch-all' provision, especially when there are more specific measures such as Article XX(b) and (g) available, not to mention the wide-ranging possibilities of the SPS and TBT Agreements. And although adjudicating on the moral considerability of domestically felt issues is very difficult, the WTO will in any event require its members to establish the domestic moral considerability of both the issue and the specific measures.[24] As ever, this raises difficult questions of evidence, but a combination of argument, information on context, and evidence of public views could be used to demonstrate that the measure aims at issues of public morality.

The objectives pursued in the regulation of GMOs are blurred and complex, and, as one claimed purpose among many, Article XX(a) risks diverting attention from more easily grasped and conceptualised social objectives. As in the EU context, notwithstanding the theoretical openness of the legal framework, it is not terribly surprising that governments tend to keep quiet about any moral objectives of regulation.

---

22 *Shrimp/Turtle*, above n. 10, para. 129.

23 Gareth Davies, 'Morality Clauses and Decision-Making in Situations of Scientific Uncertainty: The Case of GMOs' (2006) *Hebrew University International Law Research Paper No. 10-06,* argues that dealing with unknowns fits in the morality proviso (rather than in a discussion of science and regulatory autonomy as argued here), on the basis that the concern is not the environmental harm, but the 'risk' of environmental harm.

24 See also Jeremy C. Marwell, 'Trade and Morality: the WTO Public Morals Exception after Gambling' (2006) 81 *New York University Law Review* 802.

Whilst particular social values are allowed into WTO debate by the lettered paragraphs of Article XX, they can be applied only within certain boundaries. The validity of social objectives is confined by the 'chapeau' of Article XX, which prohibits the application of measures 'in a manner which would constitute a means of arbitrary or unjustifiable discrimination between countries where the same conditions prevail, or a disguised restriction on international trade.' The US demand for turtle excluder devices, the Brazilian ban on retreaded tyre imports and the US restrictions on internet gambling[25] all fell foul of the chapeau.

Shrimp/Turtle is the most detailed Appellate Body examination of the chapeau. The chapeau demands the identification of 'a line of equilibrium between the right of a Member to invoke an exception under Article XX and the rights of the other members'.[26] The obligation is one of good faith, and importantly the Appellate Body imposed no *explicit* hierarchy between the values (non-discrimination/international trade and protection of turtles). The Appellate Body characterised the US measures as 'rigid and unbending', especially in requiring a specified technique of turtle protection across the board.[27] And, because the determinative factor was whether the country in whose water the shrimp were caught has been certified by the United States, even shrimp caught by the allowed methods might be excluded from the US market. The measures were applied at the border through procedures characterised as 'singularly informal and casual', amounting to a denial of 'basic fairness and due process'.[28]

This decision is cited as the beginnings of a procedural approach to WTO disciplines, whereby the WTO concentrates on due process and fairness rather than trying to assess the substance of domestic regulatory choices. The relationship between due process and the substantive question of 'arbitrary and unjustifiable discrimination' is unclear, not least because 'due process' sometimes fits uncomfortably in the 'discrimination' rubric: it must be as bad to treat all with no due process as some.[29] But a due process approach to WTO obligations has been welcomed, since it demands consideration of the situa-

---

[25] The gambling measures were dealt with very briefly by the Appellate Body and condemned because of discriminatory application between domestic and foreign service providers, para. 351.

[26] *Shrimp/Turtle*, above n. 10, para. 159.

[27] Sanford E. Gaines, 'The WTO's Reading of the GATT Article XX Chapeau: A Disguised Restriction on Environmental Measures' (2001) 22 *University of Pennsylvania Journal of International Economic Law* 739.

[28] *Shrimp/Turtle*, above n. 10, para. 181.

[29] See de Búrca and Scott, above n. 2.

tion of outsiders, whilst respecting regulatory autonomy.[30] Similarly, the apparent encouragement of international negotiation on matters of common interest (the negotiation of multilateral agreements with some and not others is one of the areas of discrimination; see further below p. 226) has been welcomed. The approach in *Shrimp/Turtle* is, however, not without its critics. First, the due process approach could be oppressive, with its own normative implications as to the proper way of making decisions, and it could impose 'extraordinary preconditions' on recourse to Article XX.[31] Another, opposite, concern is that, in the absence of some substantive vigilance, due process can become a mere box-ticking exercise, allowing members with plenty of administrative resources to escape their free-trade undertakings. The *Brazil–Tyres* focus on the purposes of discrimination could be a useful supplement in this respect. Discrimination existed in the non-application of the ban to MERCOSUR countries, and in the continued import of used tyres (which would then be retreaded in Brazil) under individual court injunctions. The Appellate Body examined the rationale provided for the discrimination, which must itself relate to the pursuit of the Article XX objectives (or at least 'not go against the objective that was provisionally found to justify a measure under a paragraph of Article XX') if it is to escape the chapeau.[32] Whatever the convenience and advantages of procedural approaches, it is clear both that they can be very demanding and that the Appellate Body does ultimately engage with the substance of national decisions. The normative impact of the WTO cannot be avoided, whether through substantive or procedural approaches.

The 'chapeau' of Article XX applies to all the exceptions, and hence imposes certain constraints on all the social values pursued via Article XX. An additional constraint on the application of measures to protect health under paragraph (b) or public morals under paragraph (a) is that the measure be 'necessary'. The demands of 'necessity' are still not entirely clear.[33]

---

30 See for example Armin von Bogdandy, 'Law and Politics in the WTO: Strategies to Cope with a Deficient Relationship' (2001) 5 *Max Planck Yearbook of United Nations Law* 609, who argues for a procedural approach in cases of ambiguity.

31 Gaines, above n. 27, p. 745.

32 *Brazil – Measures Affecting Imports of Retreaded Tyres* (WT/DS332/AB/R, 3 December 2007), para. 227. The Appellate Body rejected the Panel's approach, which was to assess the *effect* of the discrimination (whether the objectives were being 'significantly undermined', para. 7.306). The effect of the discrimination may however be a relevant factor, para. 230.

33 See Jan Neumann and Elisabeth Turk, 'Necessity Revisited: Proportionality in World Trade Organization Law after *Korea–Beef, EC–Asbestos and EC–Sardines*' (2003) 37 *Journal of World Trade* 199. 'Relating to' is the requirement under Article XX(g).

Necessity was considered in detail by the Appellate Body in the *Asbestos* dispute, which concerned a complaint from Canada about a French ban on asbestos products. The legitimate objective of the measure was the protection of human health (Article XX(b)). This much was uncontroversial. The key legal dispute was the 'necessity' of the measure. Canada argued that a less restrictive approach, specifically controlling use of asbestos rather than banning it, would suffice. The Appellate Body confirmed that WTO members have the right to determine their appropriate level of protection of health, and identified France's chosen level of health protection as a 'halt' to the spread of asbestos-related health risks.[34] France was not obliged to pursue an alternative measure that would not 'halt', but just reduce, the disease. Brazil's objectives in its import ban on retreaded tyres was 'the reduction of the risks associated with waste tyre accumulation to the maximum extent possible',[35] and, as in *Asbestos,* alternatives proposed by the complainant, including a range of improved management options (landfill, incineration, recycling, for example) did not meet this standard. According to the Appellate Body, 'non-generation measures are more apt to achieve this objective because they prevent the accumulation of waste tyres, whilst waste management measures dispose of waste tyres only once they have accumulated'.[36]

Necessity implies at least the assessment of whether less trade-restrictive measures are 'reasonably available' to meet the same objective. A measure is not 'reasonably available' if 'it is merely theoretical in nature, for instance, where the responding Member is not capable of taking it, or where the measure imposes an undue burden on that Member, such as prohibitive costs or substantial technical difficulties'.[37] And at this stage the level of protection is not up for grabs. There should be no balancing of ends and means – a member can apparently seek to eliminate an identified risk even if a slightly lower level of protection would be dramatically less trade restrictive. But a high level of protection does not provide carte blanche. The Appellate Body in *Brazil–Tyres* explicitly disagreed with Brazil's argument that 'because it aims to reduce risk exposure to the maximum extent possible, an import ban that brings a marginal or insignificant contribution can nevertheless be considered necessary'.[38] Brazil had to establish not only the existence of the risk but also the material contribution of the challenged measure to addressing that risk. A qualitative rather than a quantitative assessment of those issues sufficed,

---

34 *EC–Asbestos*, above n. 6, para. 168.

35 *Brazil–Tyres*, above n. 32, para. 144.

36 *Brazil–Tyres*, above n. 32, para. 174. The risks associated with the proposed alternatives are also relevant.

37 *Gambling*, above n. 16, para. 308.

38 *Brazil–Tyres*, above n. 32, para. 150.

however.[39] The evidential requirements are not apparently particularly burdensome, which is worth bearing in mind when we return to the SPS Agreement below.

Necessity also, however, according to the Appellate Body, requires a 'weighing and balancing',[40] which potentially allows for more intensive review: ' "[t]he more vital or important [the] common interests or value" pursued, the easier it would be to accept as "necessary" measures designed to achieve those ends'.[41] In *Asbestos*, the value pursued was 'the preservation of human life and health through the elimination, or reduction, of the well-known, and life-threatening, health risks posed by asbestos fibres' and this value was held to be 'both vital and important in the highest degree'.[42] How *other* objectives will measure up is not clear. In *Brazil–Tyres*, whilst the protection of human health from mosquito-borne disease and exposure to toxic emissions from tyre fires was found by the Panel to be 'both vital and important in the highest degree', the protection of the environment was 'important'.[43] Even asking trade panels to 'rank' values is problematic, but the precise impact of identifying an objective as less than 'vital and important in the highest degree' is not clear. If the allocation of a lesser status to certain non-economic values pursued in domestic regulation requires the member to forfeit its chosen level of protection in the interest of trade, the intrusiveness of WTO disciplines is enormously enhanced. The range of objectives pursued in EC regulation of GMOs makes this a particularly sensitive matter. But the Appellate Body seems to be conscious of its shaky regulatory legitimacy and

---

39 *Brazil–Tyres*, above n. 32, paras 151–2.

40 It is also of course possible that balancing might enter the equation at a different point of the WTO legal analysis. In *Japan – Measures Affecting the Importation of Apples* (WT/DS245/AB/R, 26 November 2003), for example, the 'clearly disproportionate' nature of measures taken against a 'negligible' risk was said to constitute evidence (only) to the effect that there is no rational relationship between the scientific evidence and the measures taken under the SPS Agreement, paras 163–8, in respect of Article 2.2 SPS Agreement. The appropriate approach depends on the particular case, especially para. 164.

41 *EC–Asbestos*, above n. 6, para. 172, citing *Korea – Measures Affecting Imports of Fresh, Chilled and Frozen Beef* (WT/DS161, WT/DS 169/AB/R, 11 December 2000). *Brazil–Tyres*, above n. 32, para. 178.

42 *EC–Asbestos*, above n. 6, para. 172.

43 *Brazil–Tyres*, above n. 15, para. 7.111–112 (Panel Report), Appellate Body, above n. 32, para. 172. At para. 144, the Appellate Body quotes the Panel's reference to Brazil's argument that 'few interests are more vital and important than protecting human beings from health risks, and that protecting the environment is no less important'. The Panel was noting Brazil's arguments, not explicitly adopting them, but it is interesting that the Appellate Body had no desire to emphasise the different levels of importance of different objectives.

has consistently emphasised the importance of allowing members to choose their own level of protection. The Appellate Body does not seem actually to engage in any weighing and balancing in *Asbestos*, although in *Brazil–Tyres* it rejected that very argument from the EC, which argued that the Panel had not actually *done* the weighing and balancing.[44] The weighing and balancing is a 'holistic operation that involves putting all the variables of the equation together and evaluating them in relation to each other'.[45] But the Appellate Body does not demand a particularly intrusive review from Panels on this element of 'necessity'.

The SPS Agreement is designed to 'elaborate rules' for the application of the GATT, particularly Article XX(b), although it applies independently of any breach of the GATT. The objectives open to discussion under the SPS Agreement are found in the definition of an SPS measure in Annex A, essentially (like Article XX(b)) human, animal and plant life and health, but (unlike Article XX(b)), specifically as affected by particular risks, for example plant-borne disease or contaminated food. The SPS Agreement also addresses broader 'other damage' caused specifically by the entry, establishment or spread of pests.[46] The definition of an SPS measure was surprisingly broadly applied by the *EC–Biotech* Panel, as discussed below (pp. 234–6). Whilst important social values of (basically) human, animal and plant safety are thoroughly embraced within the SPS Agreement, this is in a context highly constrained by scientific disciplines, explored below (pp. 211–22).

The TBT Agreement expands the range of social values relevant in the WTO, but again subject to constraints. Technical regulations must neither create 'unnecessary obstacles to international trade' nor be 'more trade restrictive than necessary to fulfil a legitimate objective'.[47] Necessity, as discussed above, has something of a history in WTO law, and allows review at least of

44 *Brazil–Tyres*, above n. 32, para. 176.

45 *Brazil–Tyres*, above n. 32, para. 182.

46 The SPS Agreement defines an SPS measure as: Any measure applied:

(a) to protect animal or plant life or health within the territory of the Member from risks arising from the entry, establishment or spread of pests, diseases, disease-carrying organisms or disease-causing organisms;

(b) to protect human or animal life or health within the territory of the Member from risks arising from additives, contaminants, toxins or disease-causing organisms in foods, beverages or feedstuffs;

(c) to protect human life or health within the territory of the Member from risks arising from diseases carried by animals, plants or products thereof, or from the entry, establishment or spread of pests; or

(d) to prevent or limit other damage within the territory of the Member from the entry, establishment or spread of pests.

47 TBT Agreement, Article 2.2.

whether a less trade-restrictive option is available, and also potentially more intrusive 'balancing' approaches to proportionality. The TBT Agreement defines the 'legitimate objectives' of a technical regulation *non-exhaustively* as 'national security requirements; the prevention of deceptive practices; protection of human health or safety, animal or plant life or health, or the environment'.[48] This is a much broader range of legitimate objectives than is found elsewhere in the GATT and related agreements, and the only explicit reference to environmental protection. It also leaves space for discussion of other non-specified 'legitimate objectives', which could include 'new' or unexpected values. The Appellate Body has allowed the legitimacy of 'market transparency, consumer protection and fair competition' through this provision.[49]

A Panel is able to carry out 'an examination and a determination on the legitimacy of the objectives' pursued under the TBT Agreement.[50] As with the second-guessing of moral exceptions, however, it would be problematic for the WTO to condemn as 'illegitimate' any social (or religious or ethical) objective genuinely pursued by a member on behalf of its people, except of course economic protectionism. Objectives that fall somewhere *between* social values and protectionism are particularly difficult. Protecting organic agriculture (perhaps by strict coexistence rules that make certain areas de facto GM-free) is the obvious example, as also in the EU (see Chapter 4). A range of social goods is provided by organic agriculture, but pursuing those social goods *by means of* the protection of organic agriculture is arguably also protection of a particular domestic form of production. The protection of any traditional industry from ruin by new technology may imply both social values and economic protection. But these social values are difficult to distinguish from the obvious sacrifices made for the benefits of international trade, as witnessed perhaps by the rapid decline of traditional heavy industry in western Europe over recent decades. Organic agriculture merits protection, but there are many incentives on members to express their regulatory objectives modestly.

Determining the purpose of a measure is central to its assessment under WTO disciplines. The *EC–Biotech* Panel examined both the stated aim of the legislation and the measure itself. It went beyond 'the subjective intent of the legislators or regulators'.[51] So, for example, Austria claimed that its safeguard measures rested in part on concern about lack of consumer information. The Panel, however, observed that the safeguard clauses provided in the EU

---

48 Article 2.2.

49 *European Communities – Trade Description of Sardines* (WT/DS231/AB/R, 26 September 2002), para. 263. This was not disputed by the complainant.

50 *Sardines*, above n. 49, para. 286.

51 *EC–Biotech*, above n. 4, para. 7.2558.

legislation (discussed in Chapter 3) do not allow that purpose.[52] The Panel accordingly interpreted Austria's labelling measures in terms of unanticipated adverse effects on human health, bringing Austria within the SPS Agreement and extending EU restrictions on national autonomy into the international arena. The Panel took a more generous approach to the assessment of a single measure that pursues multiple objectives. To the extent that a measure pursues an SPS purpose, it is assessed under the SPS Agreement, but to the extent that the same measure pursues a non-SPS objective, it can be assessed separately under a different WTO Agreement. So a measure that is an SPS measure might also (and independently) be a TBT measure. This in turn should mean, for example, that a measure that does not comply with the SPS Agreement in its efforts to protect human health may nevertheless pursue legitimate consumer protection objectives, justifiable under the TBT Agreement. This provides some relief from the expansion of SPS disciplines in *EC–Biotech*, discussed below (pp. 234–6). It requires, though, that the non-SPS objective is capable independently of justifying the measure, of providing 'an autonomous raison d'être, i.e., a different purpose which would provide an independent basis for imposing the requirement'.[53] The Panel did not go on to consider other objectives and other Agreements in *EC–Biotech*.

WTO members are well used to bringing their measures within a framework of particular allowed objectives. The agreements are potentially broad in the social values they anticipate, especially the 'morality' provision in the GATT and the open-ended nature of the TBT Agreement's 'legitimate objectives'. It is nevertheless likely that certain social values will be more readily heard than others – so, for example, the protection of human life and safety is clearly not just legitimate, but 'vital and important in the highest degree'. There are no such guarantees that restrictions on GMOs about, for example, protecting the sensitivities of religious conservatives would find space in the framework of the TBT Agreement or the 'public morals' defence. It is potentially even more difficult to respond to, for example, distributive concerns about corporate control of the food sector. A conservative rationalisation of measures has plain attractions.

Simple response to public concern is also difficult within the WTO framework, vulnerable to charges both that it indulges public prejudices and that public prejudices are being manipulated with protectionist ends in view. Frustratingly, although perhaps wisely, the Appellate Body has said very little on the democratic responsiveness of domestic decision making. It came closest in *Hormones*, in which the EC's ban on the import of beef from cattle

52 *EC–Biotech*, above n. 4, para. 7.2647.
53 *EC–Biotech*, above n. 4, para. 7.165. See Scott, above n. 8, pp. 19–21.

reared using certain hormones was challenged. The Appellate Body, whilst condemning the ban as (broadly) inadequately supported by scientific evidence, also worked hard to recognise some of the difficulties of requiring a scientific basis for regulation. Some of the Appellate Body's openness on science in this decision rests on what 'responsible and representative governments' do. So, for example:

> In most cases, responsible and representative governments tend to base their legislative and administrative measures on 'mainstream' scientific opinion. In other cases, equally responsible and representative governments may act in good faith on the basis of what, at a given time, may be a divergent opinion coming from qualified and respected sources.[54]

And similarly, when deciding on the sufficiency of the scientific evidence on which a member claims to base its SPS measures, a Panel should bear in mind 'that responsible, representative governments commonly act from perspectives of prudence and precaution where risks of irreversible, e.g. life-terminating, damage to human health are concerned'.[55] The Appellate Body also examined the EC measure under Article 5.5 SPS Agreement, which seeks 'consistency' in the application of SPS measures in different situations.[56] The challenge was essentially that different hormones with similar impacts were regulated differently. The Appellate Body decided that the differences in regulation did not lead to 'discrimination or a disguised restriction on international trade', in part because of evidence on the 'depth and extent of the anxieties experienced within the European Communities'. The mere fact of 'supposed multiple objectives' (which as well as safety included harmonisation and consumer reassurance) pursued by the EC did not seem to trouble the Appellate Body as it had the Panel.[57] The potential for consumer, and apparently also European Parliament,[58] concerns to explain the non-protectionist nature of different treatment of otherwise similar risks is an important restriction on the reach of Article 5.5, and is indicative of the general effort of the Appellate Body in *Hormones* to recognise the political as well as technical complexity of risk regulation.

---

54 *EC – Measures Concerning Meat and Meat Products* (WT/DS26/AB/R, WT/DS48/AB/R, 16 January 1998), para. 194.

55 *Hormones,* above n. 54, para. 124.

56 For detailed discussion, see Scott, above n. 8, Chapter 4.

57 *Hormones,* above n. 54, para. 245. *Australia – Measures Affecting Importation of Salmon* (WT/DS18/AB/R, 20 October 1998) seemed to take a more objective approach; see Scott above n. 8, Chapter 4.

58 Which the Panel had been concerned about, but the Appellate Body did not return to, *Hormones,* above n. 54, paras 243–5.

Similarly, public anger about GMOs was obliquely discussed in *EC–Biotech*. The Panel acknowledged that 'anticipated member State opposition' was one reason for the delay by the Commission, and rightly acknowledged concern for 'the legitimacy and acceptability of an eventual decision by the Commission' in the face of such opposition.[59] This apparently rather compelling concern about the legitimacy of decision making did not however justify delay:

> Were it otherwise, the obligation to complete approval procedures without undue delay would impose no real discipline as the Commission could then suspend approval procedures every time it anticipates significant member State opposition and regardless of whether there are valid reasons for such opposition.[60]

This rather casual reference to the 'validity' of political opposition is important. It hints that the reasons for opposition have to be couched in WTO terms, a sort of subsidiary approach to democratic demands that we also see in *Hormones*: responsiveness, but responsiveness in the exercise of a specified WTO discipline. To echo the debate, public concerns and public values must be fitted into a tight WTO framework of legitimacy and constraints. Most public concern, if properly examined, should be capable of being framed in terms of the categories allowed in the agreements. But again, some will be easier to argue than others, and the relevance of the nature and importance of the value pursued (in *Hormones*, protection from 'life-terminating' harms) is not clear.

The WTO faces the huge challenge of allowing the domestic pursuit of non-economic values without reducing confidence in the reciprocity of trade values, and without inciting retaliation or a loss of faith in the system. The EU clearly had its own long-term institutional legitimacy in mind when it ceased to push ahead with authorisation applications for GMOs, a lesson that the WTO should heed. The WTO agreements do provide generous space for the consideration of a very wide range of social objectives. However, social values pursued at a domestic level have to compete with the dominant objectives of trade liberalisation, and some of those social values are likely to be heard more clearly than others.

SOCIAL VALUES AND DEMOCRATIC VALUES

WTO disciplines challenge traditional understandings of democratic responsibility within the state (or EU), putting barriers in the way of responses to

---

59 *EC–Biotech*, above n. 4, para. 7.1808.
60 *EC–Biotech*, above n. 4, para. 7.1808.

democratic demands. But this concern is matched by convincing arguments that the WTO can exercise a democracy-enhancing impact on domestic regulators, forcing them to respond to concerns from beyond their immediate constituency. National democratic pressures rarely consider properly the impacts of a decision on geographically distant producers and societies, and when they do (for example, concern about environmental or labour standards in developing countries) they may do so without a proper understanding of the local context. US claims that EU suspicion of GMOs contributed to hunger in Southern Africa during the 2002 food shortages are discussed in Chapter 2. Whatever the truth or convenience of that particular claim, it reminds us that the 'feed the world through biotech' agenda might resonate less urgently in the well-fed EU than elsewhere. Whether precaution lies in pushing ahead with biotechnology or in stopping will involve different calculations from different perspectives.[61] Forcing powerful economies (such as the EU) to consider the burden they put on the less powerful when regulating to protect or indulge their own publics has obvious attractions. Even with respect to relations among the powerful, it is difficult not to criticise the application of the EU's moratorium. It was informal, with no effort at all to consider the impact on those affected or to explain the situation formally to them, and no opportunities for appeal – in short, no accountability at all.

Requiring some consideration of outsider perspectives, through formal consultation obligations in both the SPS and the TBT Agreements (below p. 207), more ambiguous 'due process' demands or simply a knowledge that regulation may need to be explained and defended, has the potential to provide an important alternative perspective on the WTO. And even in terms of internal responsiveness, at their best the constraints imposed by WTO membership can expose members to 'more and better' information, inducing 'a healthy destabilisation of established patterns of behaviour and underlying default assumptions'.[62] Although there are serious doubts about the ability of external

---

61 See Cass R. Sunstein's discussion of the 'narrow viewscreen' we use when assessing caution, 'Beyond the Precautionary Principle' (2003) 151 *University of Pennsylvania Law Review* 1003. See Chapter 2 (p. 45). Scott, above n. 8, discusses the impact of EU regulation on freshwater fishing in Kenya, pp. 41–4. Dale D. Murphy, 'The Tuna–Dolphin Wars' (2006) 40 *Journal of World Trade* 597 discusses some of the human and environmental consequences of 'dolphin-friendly' tuna regulations in the US – regulations actively sought by the dominant US corporations in the industry, and famously the subject of GATT disapproval.

62 Scott, above n. 8, in the context of the SPS Agreement, and more generally Joanne Scott, 'European Regulation of GMOs: Thinking about Judicial Review in the WTO' (2004) 57 *Current Legal Problems* 117. See also Robert Howse, 'Democracy, Science and Free Trade: Risk Regulation on Trial at the WTO' (2000) 98 *Michigan Law Review* 2329.

accountability to compensate for the loss of domestic democratic responsiveness, a globalised world without the WTO would certainly suffer from its own legitimacy issues. One thing that might affect our view of the trade-off between internal and external responsiveness is the question of to whom or what interests members are most alert in the other-regarding exercise of WTO disciplines. The costs to domestic democratic responsiveness of forcing the EU to consider the needs of poor African farmers, for example, may be more palatable than a shift of attention from domestic citizens to the needs of transnational corporations.

It is obviously difficult to be sure what is going on, but patterns of dispute settlement might give us some small insight. Only states can be parties to a WTO dispute, and powerful economies have more frequent and successful recourse to dispute settlement. The EC and the US dominate as complainants, defendants and third parties, and developing countries are conspicuously absent.[63] Not only do small economies lack the institutional and financial resources to bring claims, but the smaller value of their trade makes the costs of participation less worthwhile, even if the effect of another member's trade measures internally is immense. In addition, the nature of remedies in the WTO, that is, trade retaliation, is discouraging for countries with weak economies. The country breaching the WTO rule is not especially bothered by the threat of retaliation (it is likely that the US is more attuned to the views of the EU than, say, Antigua), and a developing country may impose significant costs, which it can least afford, on *itself* in undertaking trade retaliation. The big economies are 'repeat players' in the WTO dispute settlement system.[64] Repeat players are able to pursue strategic litigation, looking not only for success in a specific case but also for a trade framework interpreted in a way that best reflects their interests. Quite how best to reflect those interests (any state can find itself on any side of a trade dispute) is not always going to be obvious, but the imbalance of participation in dispute settlement is nevertheless a matter of concern. The ability to litigate provides potential to shape the development of the law through formal dispute settlement, and is particularly

---

63 See the statistics in Gregory Shaffer, 'Recognizing Public Goods in WTO Dispute Settlement: Who Participates? Who Decides?' (2004) 7 *Journal of International Economic Law* 459; for discussion in the SPS context, see Scott, above n. 8, pp. 307 and onwards.

64 In domestic litigation, see Marc Galanter, 'Why the "Haves" Come Out Ahead: Speculations on the Limits of Legal Change' (1974) 9 *Law and Society* 95. There is a huge debate about whether the 'haves' always have or still come out ahead (for a flavour of the debate, see Herbert M. Kritzer and Susan Silbey, *In Litigation: Do the Haves Still Come Out Ahead?* (Stanford University Press, 2003)), but the potential structural advantages for big economies in the WTO are clear.

important in the WTO, given the virtual impossibility of amending decisions through political processes.[65]

Joanne Scott convincingly argues, however, that in considering the power dynamics of the WTO we need to look not just at the quasi-judicial, top-down lawyerly work of panels and the Appellate Body, but also beyond to the 'tentacular committee system'.[66] A number of committees operate within the WTO family. Most significantly for current purposes, both the TBT and the SPS Agreements require members to provide notice of proposed measures and an opportunity for consultation. When it works well, notification and consultation through committee operates in advance of domestic implementation of a measure, in a less hierarchical, more cooperative way than dispute settlement, and allows negotiation and adjustment of measures rather than the all or nothing of 'litigation'. And not only do developing countries seem to be more active in committees than in litigation, but enhanced regulatory standards (or wider application of higher standards) seem to be as prominent as the deregulatory agenda more commonly associated with the WTO.[67] This is of course hugely complex, but it is important to bear in mind that there are other fora, beyond panels and the Appellate Body, for debate on social regulation in the WTO. Scott's important observation on 'the power of face-to-face interaction in inducing cooperation between states'[68] challenges some of our assumptions about the WTO. It fits well with a demanding approach to due process for regulation that affects trade and with the other-regarding obligations of the WTO, providing a forum for consideration of otherwise marginalised interests. The committee system also looks very much like another piece of the multi-level governance jigsaw common at EU level. As ever with multi-level governance, there is some uncertainty about who and what interests are cooperating. And, as with multi-level governance in the EU, it works best when resolution through compromise or persuasion is possible. When there is no consensus, there is always the possibility of returning to hierarchy, because the disappointed objector either gives up or turns to the formal arena of the Dispute Settlement Understanding. The debate in committee takes place 'in the shadow of the law', including the way that law is shaped, or might be shaped,

---

[65] See von Bogdandy's discussion of the missing legislator, above n. 30. The same point applies to committees.

[66] Scott, above n. 8, p. 45. Committees also carry out important interpretative work, although the legal status of their decisions is not entirely clear. Joseph Murphy and Les Levidow, *Governing the Transatlantic Conflict over Agricultural Biotechnology: Contending Coalitions, Trade Liberalisation and Standard Setting* (Routledge, 2006), discuss the networks and committees operating on GMOs outside the WTO.

[67] Scott, above n. 8.

[68] Scott, above n. 8, p. 75.

by litigation. This is not to deny the potential of these institutional innovations to escape barriers to international regulatory accommodation. These innovative approaches cannot properly be examined by fixing them 'beneath' tried and tested hierarchical law, not least because the old approaches (here, litigation) may just not be workable at all or desirable to *any* of the players.[69] But, nevertheless, the institutional option of litigation remains, especially for more powerful members, and the hierarchy (the Panels and the Appellate Body) have to respond if one of the parties demands it.

It seems clear that the priorities of the world's larger economies are well represented in formal dispute settlement. A further, and more difficult, question is what interests shape those priorities. The democracy-enhancing potential of the WTO Agreement depends at least in part on the responsiveness of governments in the use of their WTO prerogatives. The big corporations are likely to have greater access to the WTO through their governments because of their financial and intellectual (expertise, information, lobbying) resources,[70] especially if what governments are doing is not very transparent to the broader domestic constituency. Some public interest groups are also likely to follow certain cases carefully. Whilst the involvement of corporations and interest groups may have knock-on benefits for some consumers or workers, the most vulnerable groups, on whose behalf we might be most willing to consider compromising internal democratic responsiveness, are not likely to be present in dispute settlement proceedings. Even the public interest groups are more likely to be 'northern' NGOs with northern priorities and northern funding,[71] although of course some make genuine efforts to step beyond this, and there are many examples of collaboration.[72] Southern NGOs face the same resource limitations as their governments, which means that increased attention to outsiders runs the risk of further emphasising the concerns of the wealthy 'north'.

---

[69] See Charles F. Sabel and Jonathan Zeitlin, 'Learning from Difference: The New Architecture of Experimentalist Governance in the European Union', *European Governance Papers* C-97-02. Looking at government regulation rather than litigation, Sabel and Zeitlin reject the idea that 'new governance' techniques operate in the shadow of 'government' hierarchy.

[70] See for example Susan Sell's analysis of the development of the TRIPS Agreement, *Private Power, Public Law: The Globalization of Intellectual Property Rights* (Cambridge University Press, 2003).

[71] See Gregory Shaffer, 'The World Trade Organization under Challenge: Democracy and the Law and Politics of the WTO's Treatment of Trade and Environment Matters' (2001) 25 *Harvard Environmental Law Review* 1, in the context of the WTO's political bodies.

[72] For example in the amicus briefs in *EC–Biotech*; see n. 74 below.

The nature of WTO discourse compounds the difficulty of participation in dispute resolution. The approach of panels and the Appellate Body is highly specialised, case and context specific, often lengthy and complex. The *EC–Biotech* Panel decision, for example, is almost overwhelmingly lengthy at 1200 pages, plus a large set of annexes. The length and complexity of decisions not only increases the cost of state participation but also deters broad external scrutiny. The *EC–Biotech* decision, like all Panel decisions, has profound political as well as legal significance, but is likely to be read in its totality, let alone understood in all its subtlety, only by trade specialists, and by very few of them. The opacity of this decision has a real impact on the ability of those beyond the elite to make their views resonate within the WTO.

The *EC–Biotech* Report is in fact an especially complicated (and depressing) story on transparency. The interim report, which is supposed to be confidential,[73] was very widely leaked and discussed. It is perhaps churlish to complain about brevity, but in a somewhat petulant response to those who had 'inadvertently or on purpose . . . misconstrued' its findings, the Panel appended the remarkable three-page Annex K to this 1200-page document between the circulation of the leaked draft and the final report. The Panel's dismissive tone in Annex K towards 'groups and members of civil society' that have dared to 'discuss and analyse' its interim decision bodes rather ill for a future of deliberative democracy in the WTO. Its explanation of where these 'groups' have got it wrong does not even begin to engage with their concerns. The Panel may well have known what it meant, but, as a quasi-judicial body, what it means has to be equally clear to everybody else. The resistance to outsider perspectives could not be clearer. In addition, whilst the Panel exercised its discretion to accept the three amicus briefs submitted to it, 'in rendering our decision, we did not find it necessary to take the *amicus curiae* briefs into account'.[74] And that is it. There is no explanation of this conclusion, let

---

[73] It is reasonable for the Panel to be concerned by the leak – but not to direct its anger at the 'groups' that disagree with it. And substantively, whilst the Panel in Annex K correctly emphasises the space it leaves to domestic regulators, it remains the case, and unacknowledged, that that space is tight. Note that, according to the Panel, the Annexes 'form an integral part' of its Report, para. 7.34.

[74] *EC–Biotech*, above n. 4, para. 7.11. The first *amicus curiae* brief was from a group of academics in the field of sociology of risk (Lawrence Busch, Robin Grove-White, Sheila Jasanoff, David Winickoff and Brian Wynne, also the authors of David Winickoff et al., 'Adjudicating the GM Food Wars: Science, Risk and Democracy in World Trade Law' (2005) *Yale Journal of International Law* 81); the second from a group of non-governmental organisations represented by the Foundation for International Environmental Law and Development (FIELD, based in the UK) and including developing country groups; the third from a group of US environmental and consumer interest groups represented by the Center for International Environmental Law, para. 7.10.

alone any engagement with the arguments made by civil society. This is not a one-off: the Appellate Body in *Asbestos* for the first time invited outsiders to apply to submit a brief on the dispute, but then denied leave to all of the applications without giving reasons, following, in particular, resistance from developing country members.[75] This is complex, especially because some developing countries see the involvement of public interest groups concerned with the environment, labour issues or human rights as another bite of the cherry for 'northern' concerns.

But the recognition that the Appellate Body makes multi-faceted, complex and far-reaching social decisions, not just technical decisions, makes it important to reach beyond trade specialists. That need not be through amicus briefs (indeed focusing on amicus briefs, which are ad hoc and favour the well resourced, is probably an indication of just how desperate the situation is), but it is increasingly difficult to pretend that states provide adequate inclusion at the international level.[76] The dispute settlement process could provide a far more open forum for the consideration or even deliberation of trade.[77] Avoiding 1200-page reports would be a start. Hearings, pleadings and interim decisions are closed and confidential, although, in a potentially progressive development, hearings were opened to public observation for the first time in the controversial *Hormones II* dispute.[78] Just as WTO disciplines can be a useful destabilisation of old and complacent ways of domestic regulation, so might a serious engagement with outsider perspectives be constructively disruptive of dispute settlement.

Notwithstanding the difficulties, there is real value in examining the WTO as a forum for the consideration of interests otherwise excluded from domestic regulation. If the WTO bodies sometimes seem to get the balance between trade and other social objectives wrong, it is at least important to recognise that the values they seek to uphold are also important, and reach beyond economic benefits.

---

[75] See Marie-Claire Cordonnier-Segger and Markus W. Gehring, 'The WTO and Precaution: Sustainable Development Implications of the WTO Asbestos Dispute' (2003) 15 *Journal of Environmental Law* 289, especially pp. 317–19.

[76] See Howse, above n. 9, although not in the context specifically of amicus briefs.

[77] Robyn Eckersley, 'A Green Public Sphere in the WTO: The *Amicus Curiae* Interventions in the Trans-Atlantic Biotech Dispute' (2005) *Ecolomic Policy and Law: Journal of Trade and Environment Studies*, http://www.ecolomics-international.org (accessed December 2007), for example, argues that the Appellate Body could provide a forum for deliberation. The judiciary have a special role in ensuring procedural fairness, but also as disinterested decision makers who must themselves provide 'coherent, persuasive and public reasons for their decisions', p. 19.

[78] *Canada and United States – Continued Suspension of Obligations in the EC–Hormones Dispute* DS 320, 321.

RISK ASSESSMENT AS A CONSTRAINT ON REGULATION

If finding a forum for the expression of broader social, ethical, economic or political concerns about GMOs is a desperate scrabble, finding a forum for the scrutiny and discussion of the scientific assessment of the risk posed by GMOs is relatively straightforward in EU jurisdictions. The WTO, like its members, is content to allow the pursuit and scrutiny of safety, expressed through scientific evidence. But, whilst the relevance of science and risk to social decision making is relatively uncontentious among WTO members, precisely what is meant by science and risk, and precisely their role in decision making, is most certainly not. Just as science might be seen as a useful 'neutralising' tool in domestic and EU politics, in the WTO science is seen as an objective standard against which to measure regulation, preventing protectionism and aiming to balance 'the shared, but sometimes competing, interests of promoting international trade and of protecting the life and health of human beings'.[79] Whilst some condemn utterly the WTO approach to science as an epistemologically impoverished approach to the world, imposing by default a neo-liberal deregulatory agenda, others see it as the only way to keep the worst sort of politics out of regulation, for the good of all. Most of the debate is more subtle than either of these two extremes, as indeed is the approach of the Appellate Body.[80] However, the debate is a real one, with real consequences. The EU and the US do not just disagree on the appropriate role of agricultural biotechnology, but also on what 'counts' as science for legitimate regulation.

Only the SPS Agreement explicitly calls on science as a trade discipline, although science is likely to be relevant in the TBT Agreement and the GATT. For example, using Article XX(b) or (g), to protect from uncertain harms would presumably raise questions of evidence. The SPS Agreement is likely to dominate future discussion of GMOs, as indeed it did the Panel decision. Not only does the regulation of GMOs aim predominantly at protection of animal, plant and human life and health, but the Agreement has been very broadly interpreted by the *EC–Biotech* Panel (below pp. 234–5). The Panel found all of the EC's objectives to be covered by the SPS Agreement, with the exception of the objective of prevention of the misleading of consumers, and the avoidance of nutritionally disadvantageous novel foods.[81] It does matter.

79 *Hormones,* above n. 54, para. 177 (of risk assessment under the SPS Agreement).

80 See for example Scott, above n. 8; Howse, above n. 62; Winickoff et al., above n. 74.

81 *EC–Biotech*, above n. 4, para. 7.412–14. The latter sounds a little odd, but the

The application of the SPS Agreement is often disputed, simply because it contains more intrusive and rigorous provisions for the policing of domestic measures than the other WTO agreements. [82]

The formal starting point in the assessment of SPS measures is the 'basic right' of members to take 'sanitary and phytosanitary measures necessary for the protection of human, animal or plant life or health'.[83] This basic right extends to autonomy in determining the level of protection appropriate to the member,[84] including the appropriate level of tolerance of or aversion to risk.[85] The measures must be consistent with the Agreement, and 'applied only to the extent necessary to protect human, animal or plant life or health'.[86] As discussed in Chapter 3, the expressed level of protection in the EU regulatory framework for food and feed GMOs in the EU is 'a high level of protection of human life and health, animal health and welfare, environment and consumer interests', with an authorisation condition of 'no adverse effects' on human health, animal health or the environment.[87] In principle, the WTO will not undermine the autonomy of members by looking behind this standard.[88]

Whatever the level of protection chosen, members must ensure that an SPS measure is 'based on scientific principles and is not maintained without sufficient scientific evidence'.[89] According to Article 5.1, measures must be 'based on an assessment, as appropriate to the circumstances, of the risks to human, animal or plant life or health, taking into account risk assessment techniques developed by the relevant international organizations'. Whilst the Appellate

---

reasoning seems to be that nutritional differences will be compensated by a change in diet – so, for example, reduced vitamin C in oranges only affects health if most of a person's vitamin C comes from oranges.

82 Jacqueline Peel, 'A GMO by Any Other Name . . . Might be an SPS Risk!: Implications of Expanding the Scope of the WTO Sanitary and Phytosanitary Measures Agreement' (2007) 17 *European Journal of International Law* 1009; Scott, above n. 8, p. 17.

83 Article 2, heading and 2.1.

84 Article 3.3 in respect of independence from international standards, Article 4.1 in respect of the standards of other members.

85 *Hormones*, above n. 54; *Asbestos*, above n. 6; *Australia–Salmon*, above n. 57.

86 Article 2.2.

87 Regulation 1829/2003 on Genetically Modified Food and Feed [2003] OJ L 268/1, Articles 1 and 4(1).

88 Although note the discussion in Chapter 3 of how realistic this standard is. If a member does not select a level of protection (and 'no adverse effects' is clearest in respect of food and feed) the Panel will be entitled to establish the level of protection by reference to the level of protection 'reflected in the SPS measure actually applied', *Salmon*, above n. 57, para. 207.

89 'except as provided for in paragraph 7 of Article 5', Article 2.2. The Appellate Body emphasises that Articles 2.2 and 5.1 should be read together (*Hormones*, above n. 54).

Body is not shy about condemning challenged SPS measures, it has self-consciously attempted to preserve some sophistication and flexibility for domestic dealings with science. The *EC–Biotech* Panel decision is restrictive in a number of respects, but even here we see a continued *stated* commitment to domestic autonomy.

The SPS Agreement was applied in *EC–Biotech* individually to nine national safeguard measures, by six member states, and none was found to be consistent with the Article 5.1 obligation that SPS measures be 'based on' a risk assessment. To take just one (by now rather familiar) case, the complainants challenged three Austrian safeguard measures.[90] Austria argued that the relevant products had not been tested in realistic conditions and that monitoring was inadequate. It emphasised long-term effects, especially in environmentally sensitive areas, and argued that there is a risk of accelerated soil erosion and rapid loss of habitat and genetic diversity in mountain ecosystems. Austria drew the conclusion that further investigation was needed in these areas before the relevant GMOs were marketed in its territory.[91] The Panel found that Austria had failed to indicate 'relative probability of the potential risks it identifies', but instead 'makes reference to possibilities of risks or simply to the inability to determine probabilities'.[92] Relying on the Appellate Body's decision in *Australia–Salmon*,[93] the Panel stated that in the absence of an assessment of relative probability there is no adequate risk assessment. Although 'qualitative' as well as 'quantitative' assessments of probability are acceptable,[94] the mere *possibility* of risk cannot provide a basis for the Austrian measures.

This demonstrates the restrictive impact of the approach to science in the SPS Agreement. A genuine (and evidenced) *possibility* looks like a compelling reason to take action. It may be that this restrictive approach reflects concern that safety can never be proved, placing the 'possibility' of risk in the same category as 'theoretical' risk.[95] The Appellate Body has stated that 'theoretical'

---

[90] The quotations and references are taken from the Panel's assessment of Austrian measures in respect of T25 maize (unless otherwise stated), but a similar approach is taken in respect of Bt176 maize and MON810 maize.

[91] *EC–Biotech*, above n. 4, para. 7.2569.

[92] *EC–Biotech*, above n. 4, para. 7.3044.

[93] Above n. 57.

[94] And following *Hormones*, above n. 54, para. 186, and *Salmon,* above n. 57, para. 124, it need not establish a 'certain magnitude or threshold level or degree of risk'.

[95] There is, though, a close textual analysis underlying this approach. The SPS Agreement, Annex A.4 defines 'risk assessment' as 'The evaluation of the likelihood of entry, establishment or spread of a pest or disease within the territory of an importing Member according to the sanitary or phytosanitary measures which might be

risk, that is, 'the uncertainty that theoretically remains since science can never provide absolute certainty',[96] may not be addressed under Article 5.1. This is familiar from the EU's Court of First Instance decision in *Pfizer*, rejecting 'hypothetical' risk as a basis for regulation,[97] and it raises the same questions. In particular, whilst it is entirely rational that this inability to prove safety should not often be decisive, it is rather odd to rule out entirely its relevance to regulators, when perhaps the benefits of a technology are also theoretical or weak. But in the WTO there remains an option not available in the EU,[98] and that is to proceed to regulation under Article 5.7, discussed below.

With respect to questions of antibiotic resistance, allergenicity and toxicity, Austria attempted to justify its measures by identifying problems with the 'risk assessment' on which it was being asked to 'base' its measures. According to the Panel, evaluation of 'risk assessment *procedures*', rather than 'the potential for adverse effects on human or animal health arising from the consumption of specific foods containing or consisting of GMOs', is not adequate.[99] Again, one might have thought that genuine concern about existing risk assessments could provide compelling grounds for action, but apparently not. Members are entitled to legislate on the basis of a disagreement with a risk assessment, but must provide not just evidence of concern but their own study capable of meeting the definition of risk assessment: they must explain 'how and why they assess the risks differently, and . . . provide their revised or supplemental assessment of the risks'.[100] Similarly, but even more strikingly, with respect to concern about adverse effects of Bt toxin on non-target organisms, and insect resistance, the Austrian 'reasons document'

> highlights studies of undesired effects on non-target organisms related to the consumption of Bt maize but does not itself make an evaluation of the potential adverse health effects or the likelihood of these undesired effects occurring in the event that MON810 maize were to be introduced . . . the reasons document also identified one study which noted that 'further effects on the food chain [from consumption of Bt pollen by monarch butterflies] are possible'. Yet, there is no

---

applied, and of the associated potential biological and economic consequences; or the evaluation of the potential for adverse effects on human or animal health arising from the presence of additives, contaminants, toxins or disease-causing organisms in food, beverages or feedstuffs'. There are, then, two approaches to risk assessment, depending on the objective of the SPS measure. One requires the evaluation of the 'potential', the other the evaluation of the 'likelihood'.

96 *Hormones*, above n. 54, para. 186.

97 Case T-13/99 *Pfizer Animal Health SA v Council* [2002] ECR II-3305.

98 Although in the legislative context of GMOs we have 'other legitimate factors', Chapter 3, pp. 83–9.

99 *EC–Biotech*, above n. 4, para. 7.3049.

100 *EC–Biotech*, above n. 4, para. 7.3062.

> evaluation of the potential for adverse health effects or the likelihood of such effects occurring.[101]

Nor will the famous Losey study on the effect of transgenic pollen on monarch butterflies[102] suffice for the restriction of MON810 maize, because it

> focuses on a variety of Bt maize other than MON810 maize. Furthermore, while the Losey study notes that results on larvae consumption and growth rates have 'potentially profound implications for the conservation of monarch butterflies' there is no attempt to evaluate these potential implications, rather the study notes that the experimental results point to possible environmental outcomes.[103]

This passage also reminds us that risk assessment must be sufficiently specific to the risk at issue, which is also one ground on which the Appellate Body found that the EC had breached its obligations in *Hormones*. Whilst studies provided evidence that the hormones create a 'general risk of cancer',

> They do not focus on and do not address the particular kind of risk here at stake – the carcinogenic or genotoxic potential of the residues of those hormones found in meat derived from cattle to which the hormones had been administered for growth promotion purposes.[104]

The Appellate Body demands a very high degree of specificity: the conclusion that one of the 110 000 breast cancers per million women come from eating meat containing oestrogens as a growth promoter is not sufficiently specific, as it is not based on 'studies . . . focusing specifically on residues of hormones in meat from cattle fattened with such hormones'.[105] This raises a significant barrier to the regulation of small and diffuse risks. Because those risks are more or less impossible to establish with any certainty, there is a danger that they will be taken out of the range of permissible targets for regulation. The incidence of cancer from (for example) meat from hormone-treated cattle is likely to be impossible to establish (or even properly test for) *specifically*, given the high background level of disease, the diffuse nature of the risk and

---

101 *EC–Biotech*, above n. 4, para. 7.3094.

102 See Chapter 2, p. 27.

103 *EC–Biotech*, above n. 4, para. 7.3097.

104 *Hormones*, above n. 54, para. 200. See also *Japan–Apples*, above n. 40.

105 *Hormones,* above n. 54, para. 198 and footnotes. See Alan O. Sykes, 'Domestic Regulation, Sovereignty and Scientific Evidence Requirements: A Pessimistic View' (2004) 3 *Chicago Journal of Internaional Law* 353, p. 364; Joanne Scott, 'On Kith and Kine: Trade and Environment in the EU and WTO' in Joseph H.H. Weiler (ed.), *The EU, WTO and NAFTA: Towards a Common Law of International Trade* (Oxford University Press, 2000).

the small additional numbers of disease cases. And yet, scientific understanding of the carcinogenic property of the hormones, together with knowledge of their presence in the meat, might (reasonably, it would seem) lead to a wish to regulate. As Alan Sykes reminds us, placing an 'insurmountable hurdle' in the way of regulating risks 'not demonstrable through particularised scientific studies' is not a procedural restriction, but 'must surely clash with the notion that WTO law is not meant to tell Members which risks they must tolerate and which they may elect to avoid'.[106] This is extremely difficult, and Sykes himself sees the limitation as inevitable if the risk assessment obligation is to be anything more than 'window dressing'.[107] And yet it seems implausible that the Appellate Body wishes to deny even the *possibility* of regulating to protect from diseases with a high background incidence on the basis of extrapolations from similar demonstrable risks. This is just one of many unclear aspects of the WTO Agreement, and we have to hope that the case is strongly and explicitly argued when a risk assessment based specifically on extrapolation next comes before the Appellate Body.

The information provided by Austria to the *EC–Biotech* Panel was not an Article 5.1 'risk assessment', but other risk assessments were available to Austria, namely those carried out at the time of either the initial authorisation of the GMO or the subsequent EU-level consideration of the safeguard measures. These risk assessments had been used at EU level to justify *authorising* the GMO, and the question was whether safeguard measures *banning* the GMO could also be said to be 'based on' (interpreted as 'a substantive requirement that there be a rational relationship between the measure and the risk assessment'[108]) the same assessments. Much of the Panel decision leaves little room for manoeuvre on domestic approaches to science, but this is more open. One single risk assessment 'might conceivably' provide a basis for different types of SPS measures.[109] This reflects the approach of the Appellate Body in *Hormones*, which is on the whole a more optimistic basis for regulatory autonomy. According to *Hormones* the evidence from the risk assessment does not have to be 'monolithic', and minority as well as mainstream scientific views are potentially valid.[110] If a risk assessment includes a divergent opinion from 'qualified and respected sources', an SPS measure which reflects the divergent opinion can be said to be 'based on' the risk assessment.[111] This was not however established in *EC–Biotech*, and external concerns about a

---

106 Sykes, above n. 105, p. 364
107 Sykes, above n. 105, p. 368.
108 *Hormones,* above n. 54, para. 193.
109 *EC–Biotech*, above n. 4, para. 7.3064.
110 *Hormones*, above n. 54, para. 194.
111 *EC–Biotech*, above n. 4, para. 7.3060.

risk assessment which do not themselves amount to a competing 'risk assessment' are, as discussed above, not adequate as a basis for the SPS measure. Whether Austria's doubts about those risk assessments that did exist would have been sufficient for its safeguard measures had they been contained *within* the risk assessment is an interesting question – quite what the Panel expected of doubt within a risk assessment remains open. Presumably the Appellate Body is not going to allow states to draft their risk assessments strategically to allow any action, but equally the level of scrutiny does not speak for itself.

The *EC–Biotech* decision brings to the fore the more restrictive aspects of the risk assessment disciplines in the SPS Agreement. Although it is possible that the 1998 report in *Hormones* represents something of a high point in this respect, Appellate Body reports generally provide greater recognition of the sensitivity and complexity of scientific justification, and of course *EC–Biotech* has not been confirmed at that level. Some of the more open aspects of *Hormones* have already been mentioned. The most important additional factor is the broad scope that the Appellate Body attributed to the notion of 'risk assessment'. The phrase 'risk management' does not appear in the Agreement, and in introducing a supposed dichotomy between risk assessment and risk management, or scientific assessment and political judgment, the Panel in *Hormones* would have narrowed the scope of risk assessment to the scientific assessment only. The Appellate Body confirmed by contrast that a range of relevant factors can be considered in an Article 5 'risk assessment'. In *Hormones*, for example, the difficulty of enforcing good veterinary practice was a legitimate question for a SPS measure, as were differing effects depending on social and environmental context, famously expressed as the 'actual potential for adverse effects on human health in the real world where people live and work and die'.[112] So members are not restricted to basing their SPS measures on narrow laboratory studies, although, equally famously, in *Hormones* the 'real world' risk of farmers failing to observe good veterinary practice had not been subject to a sufficiently specific risk assessment.[113]

Moreover, the 'real world' cuts both ways. The *EC–Biotech* Panel apparently rejected the possibility, at least in some circumstances, of applying laboratory research to 'the real world, where people live and work and die'. So two studies of insects in the laboratory (specifically the effect of Bt maize on non-target insects) were rejected on the basis of scientific caution expressed in the studies themselves:

> the study notes that '[n]o conclusions can be drawn at this point as to how results from [...] laboratory trials might translate in the field'. This statement, in our view,

---

112 *Hormones*, above n. 54, para. 187.
113 *Hormones*, above n. 54, paras 207–8.

> implies that this study *per se* cannot be said to evaluate the alleged risks identified by Austria in its Reasons document.[114]

The industry is calling on arguments long used by those who urge caution in the application of the extrapolations and assumptions that underlie laboratory risk assessment; perhaps we should not be surprised. But we might think back to the question of relying on doubts expressed within a risk assessment to explain SPS measures (above pp. 216–17). When doubts expressed within a study about the implications of laboratory trials on 'real world' decisions put a check on regulation, they received very little scrutiny from the Panel.

It is of course impossible to assess the use of science in regulation without thinking about the precautionary principle. The Appellate Body has avoided deciding whether the precautionary principle applies in trade disputes as a principle of international law:

> the legal debate over whether the precautionary principle constitutes a recognized principle of general or customary international law is still ongoing . . . prudence suggests that we not attempt to resolve this complex issue.[115]

Notwithstanding this reluctance, the *EC–Biotech* Panel accepts that whether a member is taking a precautionary approach could 'have a bearing'[116] on the question of whether an SPS measure is 'based on' a risk assessment. In particular this could justify a different approach by different members, so that our precautionary member applies a stricter SPS measure than another member in respect of the same risk. This is a very modest approach to the precautionary principle. Whilst precaution can influence the choice of measures, the nature of the information relied upon remains the same – risk assessment under Article 5.1. Moreover, this only applies if the member is faced with a risk assessment that 'identifies uncertainties or constraints'.[117] This seems to mean that only doubt expressed within the risk assessment can justify a precautionary approach, and that precaution depends upon 'scientists' level of confidence in a risk assessment they have carried out'.[118] The sufficiency of *external* doubts is more problematic, not least because, as discussed above, the measure still needs to be 'based on' *a* risk assessment.

---

114 *EC–Biotech*, above n. 4, para. 7.3099; also 7.3098.

115 *Hormones*, above n. 54, cited in *EC–Biotech*, above n. 4, para. 7.88–89.

116 *EC–Biotech*, above n. 4, para. 7.3065. Note that *Hormones*, above n. 54, also identifies the precautionary principle in other parts of the Agreement, for example Article 3 allowing Members to choose their own level of assessment.

117 *EC–Biotech*, above n. 4, para. 7.3065.

118 *EC–Biotech*, above n. 4, para. 7.3065.

A version of the precautionary principle can, however, be found in Article 5.7:

> In cases where relevant scientific evidence is insufficient, a Member may provisionally adopt sanitary or phytosanitary measures on the basis of available pertinent information, including that from the relevant international organizations as well as from sanitary or phytosanitary measures applied by other Members. In such circumstances, Members shall seek to obtain the additional information necessary for a more objective assessment of risk and review the sanitary or phytosanitary measure accordingly within a reasonable period of time.

This 'qualified exemption'[119] to the requirement for a risk assessment comprises four cumulative elements: the scientific evidence is 'insufficient'; the measure is adopted 'on the basis of available pertinent information'; the Member which adopted the measure 'seeks to obtain the additional information necessary for a more objective assessment of risk'; and the Member which adopted the measure 'reviews the . . . measure accordingly within a reasonable period of time'.[120]

The nine safeguard measures in *EC–Biotech* all fell at the first hurdle. The question of 'insufficiency' controls the application of Article 5.7. It is also the flipside of Article 2.2, which prohibits the maintenance of SPS measures without 'sufficient scientific evidence', expressly subject to Article 5.7.[121] The Appellate Body has contrasted 'insufficient evidence' with scientific uncertainty: Article 5.7 'is triggered not by the existence of scientific uncertainty, but rather by the insufficiency of scientific evidence . . . The two concepts are not interchangeable'.[122] The *EC–Biotech* Panel was adamant that 'insufficient' includes qualitative insufficiency[123] and Article 5.7 can be invoked where the evidence is 'more than minimal in quantity, but has not led to reliable or conclusive results'.[124] We might conclude that, even if new or additional scientific information is not such as to constitute a risk assessment for

---

119 Not a defence: *EC–Biotech*, above n. 4, para. 7.2927, citing the Appellate Body. The distinction is relevant to the burden of proof.

120 *Japan – Measures Affecting Agricultural Products* (WT/DS76/AB/R, 22 February 1999), para. 89.

121 *Japan–Agricultural Products,* above n. 120, para. 80.

122 *Japan–Apples*, above n. 40, para. 184.

123 See especially Annex K. The Panel in Annex K also reminds us of the *Hormones* understanding of risk 'in human societies as they actually exist, in other words, the actual potential for adverse effects on human health in the real world where people live and work and die'. It appears that 'insufficient' evidence on these risks could open up Article 5.7, although in this case the Panel was 'not persuaded' that in the case of the safeguard measures 'the scientific evidence available at the relevant time did not allow the performance of an assessment of the risk in human societies, or natural environments, as they actually exist'.

124 *Japan–Apples*, above n. 40, para. 185.

the purposes of Article 5.1, it may unsettle the scientific information enough to make it 'insufficient', bringing Article 5.7 into play. This looks rather similar to 'uncertainty'. Perhaps the restrictive language rests again on underlying concern that the *inevitability* of scientific 'uncertainty' could open things up too far. The denial of even the relevance of 'uncertainty', however, suggests an expectation that *more* or *better* ('sufficient') science will ultimately fill in the gaps and establish the 'facts'. This wishes away the tension involved in using science as a means to justify regulatory measures in difficult cases, and takes no account of the nature of the uncertainty in respect of GMOs.

The Panel rejected the EC's argument that the question of sufficiency depends on the goals pursued by legislators. It assessed sufficiency solely according to whether the scientific evidence enabled the performance of a risk assessment for the purposes of Article 5.1,[125] binding the members closely to risk assessment under that provision. The existence of EU-level assessments from the original authorisation processes provided Austria with 'sufficient' scientific evidence for a risk assessment. And this is the case irrespective of whether that risk assessment provided the level of information or certainty that would allow Austria to apply its chosen level of SPS security. Imposing an EU-level risk assessment on EU Member States is difficult enough, further enhancing the control of the centre in EU policy. But presumably this applies in respect of risk assessments arising in other contexts (international bodies, other members, private organisations, public interest groups), placing a significant restriction on any member that thinks more or better evidence should be gathered before proceeding to the (in this case practically irreversible) adoption of a new technology. It seems perfectly understandable that what is 'sufficient' for one regulator in one social context will not be so for another. Risk assessments do not reveal universally valid facts, but are tied up with cultural and political values and assumptions, including the role of the public in decisions on new technology.[126] Where a risk is particularly sensitive, more certainty on safety might be demanded. But, according to the Panel, as long as *a* 'risk assessment' is available, SPS measures must be based on *that* risk assessment.

The *EC–Biotech* Panel seemed to see no difficulty in making recourse to Article 5.7 extraordinary. It presented Article 5.1 as the usual place for provi-

---

125 *EC–Biotech*, above n. 4, para. 7.3224. According to the Panel, the reference by the Appellate Body in *Japan–Apples* (above n. 40) to 'adequate' assessment of risks refers to whether it meets the standard and definition in the SPS Agreement, rather than one which is 'adequate for the purposes of the legislator', para. 7.3226. The Panel quotes the Codex in support, para. 7.3240. The Panel rejects the EC argument that *provisionality* is the trigger for Article 5.7, para. 7.2940.

126 See Winickoff et al., above n. 74.

sional measures, and indeed seemed willing to grant extra room under Article 5.1 to reflect the credibility of the risk assessment. So Article 5.1 allows for 'expeditious' reassessment of risks following new or additional scientific assessment, since what is 'appropriate to the circumstances' can change. But if Article 5.7 is not available there is no opportunity to rely on 'available pertinent information' rather than risk assessment. 'Available pertinent information' must be 'different in nature'[127] from the scientific evidence supporting a risk assessment under Article 5.1, but the Appellate Body has not expanded. As in the EU, quite what 'counts' as sufficient information to introduce a precautionary measure is unclear. In the EU, the information has to cross some undefined, perhaps indefinable, line between a 'purely hypothetical' approach to risk and risk which is 'adequately backed up by the scientific data available at the time'.[128] The WTO version makes no reference to scientific data, and whilst the Appellate Body has rejected recourse to 'theoretical risk' under Article 5.1, it has not considered its role in Article 5.7. Allowing theoretical risk into Article 5.7 would be a positive development, although, given that recourse to Article 5.7 is so severely limited by the *Biotech* decision, it might have little impact.

Article 5.7 allows only provisional measures, and the Member which adopted the measure must then 'seek to obtain the additional information necessary for a more objective assessment of risk' and 'review the . . . measure accordingly within a reasonable period of time'. In *Japan–Agricultural Products* the Appellate Body considered the question of 'a more objective assessment of risk', which has to go precisely to the necessity of the SPS measure at issue. And a 'reasonable period of time' is established case by case, according to the 'specific circumstances of each case, including the difficulty of obtaining the additional information necessary for the review *and* the characteristics of the provisional SPS measure'.[129]

The WTO system's demand for scientific justification of regulation is designed to ensure the objectivity of measures that affect trade. As discussed in Chapters 2 and 3, however, any attempt to take political judgment out of decisions on risk is futile, since science alone cannot resolve these questions. The greatest danger is that, if politicians and regulators are encouraged to shape their decisions around science, transparency is reduced as the real reasons for a decision are hidden from view. The Panels and the Appellate Body do recognise and assert the relevance of questions of representativeness and responsiveness in dealings with science. However, reports such as

127 *EC–Biotech*, above n. 4, para. 7.2992.
128 *Pfizer*, above n. 97, para. 144.
129 *Japan–Agricultural Products*, above n. 120, paras 92 and 93.

EC–Biotech are at best ambivalent, with professed respect for domestic autonomy and flexibility coupled with extraordinary confidence in the actual science presented. A more direct discussion of the information (cultural and social as well as scientific) that can be used to explain measures would be helpful. The most obvious place for this discussion would be under Article 5.7. But *EC–Biotech* has made recourse to Article 5.7 so difficult that grappling with this problem becomes a matter for the whole structure of the SPS Agreement.

TIME AND POLITICS

The EC argued that the *Biotech* dispute was 'a case essentially about time':

> The time allowed to a prudent government to set up and apply a process for effective risk assessment of products which are novel for its territory and ecosystems, and that have the potential of causing irreversible harm to public health and the environment.[130]

The moratorium resulted from the collapse of a regulatory structure manifestly incapable of bearing the weight of public demands. Enough Member States (taking the lead from their publics) were outraged by the flaws in the legislation to block its application until at least some concerns had been addressed. And the nature of the concern was sufficiently profound, and with sufficient impact on the commercial prospects of agricultural biotechnology, for those whom we might have expected to enforce the existing regulations (the Commission, those Member States not taking part in the moratorium, and even the industry) not to push ahead. Instead they took advantage of a breathing space, during which public concern was canvassed and regulatory approaches reviewed and renegotiated.

The Panel found that a general moratorium was in existence at the date it was established, August 2003,[131] and that the absence of a formal decision implementing the moratorium did not prevent its consideration under the SPS Agreement. The moratorium was characterised by the Panel as a 'procedural decision to delay final substantive approval decisions',[132] rather than as a

130 *EC–Biotech*, above n. 4, para. 4.500.

131 *EC–Biotech*, above n. 4, para. 7.172. The Panel did not need to express an opinion on the persistence of the moratorium beyond that date, although it did observe that the informality of the moratorium meant that it could be reintroduced at any moment, para. 7.1311.

132 *EC–Biotech*, above n. 4, para. 7.1379, 7.1382. Note the touch of irony – the Panel took over three years to give its decision.

substantive decision to reject applications. As such, it was not considered under the provisions of the SPS Agreement dealing with risk assessment. Had the complaining parties wished to rely on Article 5, they would according to the Panel have needed to argue that the underlying basis for the moratorium, that is the need to seek authorisation prior to marketing (characterised as a 'provisional ban'[133]), infringed the SPS Agreement.[134]

The Panel held that the moratorium breached the requirement in Annex C that procedures be 'undertaken and completed without undue delay'. 'Undue' is glossed as 'unjustifiable', and the key is not the objective length of time taken, but whether it can be justified on a case-by-case basis.[135] The EC's argument that the 'perceived inadequacy' of the legislative framework in 1998 could justify delay was rejected.[136] This is rather troubling. A regulatory response to public opinion on a technology inevitably follows the evolution and expression of that public opinion. Public opinion itself necessarily lags behind awareness of technology, which in turn may well follow the actual presence of the technology in the world. But the Panel implied that regulation must necessarily keep up with the commercialisation of a technology, and hence that markets set the pace.

Nor is 'evolving science, scientific complexity and uncertainty, and limited available scientific information or data' sufficient to justify delaying substantive approval decisions.[137] Rather, these issues should be taken into account in making the substantive decision. A prudent and precautionary approach is permissible under Annex C, and indeed the Panel 'perceive[d] no inherent tension between the obligation . . . to complete approval procedures without undue delay and the application of a prudent and precautionary approach'.[138] The 'core obligation' is for members to come to a substantive decision.[139] If there is 'sufficient' scientific evidence, a decision should be taken under Article 5.1; otherwise, a decision should be taken under Article 5.7:

> the SPS Agreement nowhere states that substantive decisions on applications need to give a straight yes or no answer to applicants. Members may in principle grant time-limited approvals or approvals subject to other appropriate conditions. Alternatively, they may in principle decide to reject an application subject to the

133 *EC–Biotech*, above n. 4, para. 7.1351–2.
134 *EC–Biotech*, above n. 4, para. 7.1360.
135 *EC–Biotech*, above n. 4, para. 7.1495. See Gregory Shaffer's discussion of the decision in *When Cooperation Fails: the Law and Politics of Genetically Modified Foods* (Oxford University Press, 2008, forthcoming).
136 *EC–Biotech*, above n. 4, para. 7.1511 and 7.1514.
137 *EC–Biotech*, above n. 4, para. 7. 1526.
138 *EC–Biotech*, above n. 4, para. 7.1523.
139 *EC–Biotech*, above n. 4, para. 7.1523.

> possibility of a review of that decision if and when relevant circumstances change. Relevant circumstances could include the state of scientific knowledge.[140]

The Panel categorises Annex C(1)(a) as 'essentially a good faith obligation'.[141] The EC 'delay' in the moratorium was hugely problematic in terms of process and fairness. But forcing applications through the process without delay would have ruled out one form of responsiveness to public concern, and certainly seems unlikely to have enhanced public acceptance of the technology, or indeed the long-term legitimacy of the EU. If it is not suggesting that public opinion should be ignored, or assuming (surely complacently) that it can be changed, the Panel must be suggesting that public opinion should have been dealt with in other ways. Presumably, the only realistic option open to the EU, given its difficult political situation, would have been to refuse authorisation case by case. This more open and transparent approach would have definite process advantages over the moratorium, and whilst there would be internal legal problems, the same might be said of the moratorium. But one of two things needs to be established for the EU to take this course within the rules of the WTO: either there must a risk assessment on which the EU can base each refusal of authorisation under Article 5.1; or there must be insufficient scientific evidence for a risk assessment, allowing the EU to turn to Article 5.7. This takes us straight back to the science-based disciplines discussed above. The available scientific evidence tended not, as we saw in respect of the safeguard measures, to provide a basis for denying authorisation. In the (unlikely, given the Panel's stance in *EC–Biotech* and the production of evidence by the industry) event that there is 'insufficient' scientific evidence to perform a risk assessment of some individual GMOs, provisional measures are possible under Article 5.7. But the assumption must be that, to comply with the WTO agreements as interpreted by the Panel, widespread commercialisation of GMOs in the EU would have taken place from 1998. That leaves no space at all for the public debate and negotiation of a legislative framework that took place between 1998 and 2003.

The Panel accepted that, in a situation in which science evolves and there is limited available scientific evidence, 'a deferral of substantive decisions might allow for better decisions at a later point in time', although it rejected the possibility of 'a sort of holding pattern' in the meantime.[142] Nor is a general moratorium inherently impermissible: the Panel emphasised that there may be circumstances in which it would be justifiable to impose a general moratorium. Those circumstances may include in certain cases the considera-

---

140 *EC–Biotech*, above n. 4, para. 7.1527.
141 *EC–Biotech*, above n. 4, para. 7.1498.
142 *EC–Biotech*, above n. 4, para. 7.1527.

tion of new scientific evidence that conflicts with available scientific evidence and which is directly relevant to *all* biotech products subject to a pre-marketing approval requirement. This reminds us yet again of the considerable reach of science in the SPS Agreement. The only disruption the Panel allows for is new scientific evidence, not new information, or changing social or political understandings of a problem, or the acceptability of existing science.

The *EC–Biotech* decision raises questions about the relationship between domestic democracy and international trade rules. The 'delay' in this case was not a passive refusal to act, a simple absence of action. During this period, whilst the Commission did not force through approvals of GMOs, all the institutions and Member States, and indeed outsiders including the biotech industry and environmental and consumer groups, were engaged in the difficult and prolonged negotiation of a new regulatory framework. It does not seem very realistic to suggest that pushing ahead in the teeth of opposition would have been the better solution from the perspective of even international trade, let alone legitimate decision making. It perhaps reminds us that, just as the WTO bodies are isolated from democratic pressures, they are also removed from the entire philosophy of social regulation, from the 'deeper, intuitive comprehension . . . that comes only with experience'.[143] Whilst simply allowing any response to public opinion (or democracy) could open the door to virtually any trade restrictive action, the Panel showed no awareness of the position in which EU regulators found themselves. The EU does not make a habit of abandoning its entire regulatory framework on a topic to start again from scratch. These were desperate times. To dismiss out of hand the capacity or even ambition of the EU to respond to the wishes of its citizens creates a considerably greater challenge for the WTO than to attempt to find space for delay in extremis.

INTERNATIONAL APPROACHES TO GMOs

During the moratorium, alongside the negotiation of its own regulatory framework, the EU made very serious efforts to influence the international legal framework within which WTO rules operate.[144] Many different international

---

143 Gaines, above n. 27, p. 803.

144 See Wade Jacoby and Sophie Meunier's discussion of 'managed globalisation', *Europe and the Management of Globalization: Defensive and Offensive Responses to Globalization Pressures*, http://www.princeton.edu/~smeunier/conference_europeanization.htm (accessed December 2007). The Commission's response to queries raised by other WTO members in SPS and TBT committee discussion makes this point very clearly: the document is littered with statements such as 'the European

bodies provide fora for the deliberation and negotiation of transatlantic disagreement, including bodies of the United Nations (like the Food and Agriculture Organization and the World Health Organization) and standard-setting bodies such as the Codex Alimentarius, discussed below.

As well as demonstrating a preference for collective action, there are good strategic reasons for the EU to work at international negotiation. One of the grounds for finding that the US had breached the chapeau of Article XX in *Shrimp/Turtle* was its failure to engage 'in serious, across-the-board negotiations with the objective of concluding bilateral or multilateral agreements for the protection and conservation of sea turtles'.[145] A quasi-duty to cooperate encourages a multilateral approach to collective problems. This is potentially a major limitation on the use of Article XX,[146] but could impel to the negotiating table both the state seeking to restrict trade and the state facing trade restrictions. Precisely what the Appellate Body was looking for (how serious, how across-the-board, indeed how successful) is not clear, and *Gambling* suggests that the impact of *Shrimp/Turtle* might be limited. The question of international negotiations came up in *Gambling* under the requirement that the measures taken be '*necessary* to protect public morals or to maintain public order' (above pp. 193–5). The Panel had said that failure to take up Antigua's invitation to consultation or negotiation meant that the US had not pursued a reasonably available alternative to its measures. The Appellate Body disagreed 'because consultations are by definition a process, the results of which are uncertain and therefore not capable of comparison with the measures at issue in this case'.[147] This more relaxed attitude to negotiation is rather difficult to reconcile with any very far-reaching duty-based interpretation of *Shrimp/Turtle*. It may bring the element of discrimination (the US negotiated with some states and not others) to the centre of *Shrimp/Turtle* – other countries were not involved in *Gambling*. The incentive to international cooperation is, however, even more ambiguous after *Gambling* than it was before.

Community has long supported international cooperation in this respect' and 'the European Commission is soliciting cooperation with its trading partners': for example, *Response from the Commission to Comments submitted by WTO Members under Either or Both G/TBT/N/EEC/6 and G/SPS/n/EEC/149*, G/SPS/GEN/337, G/TBT/W/179 (2002).

145 *Shrimp/Turtle*, above n. 10, para. 166.

146 Gaines, above n. 27.

147 *Gambling*, above n. 16, para. 317. Although the Panel may have meant that the failure to negotiate provides evidence of failure to examine reasonably available alternative measures, rather than seeing the negotiation itself as an alternative, Eric H. Leroux, 'Eleven Years of GATS Case Law: What Have We Learned?' (2007) 10 *Journal of International Economic Law* 749.

The most high-profile international intervention on GMOs is the Cartagena Protocol on Biosafety, slowly and painfully negotiated under the auspices of the UN Convention on Biological Diversity.[148] More than 140 countries are parties to the Protocol, although some of the main GMO-cultivating countries are not, including the US. The Cartagena Protocol rests on the principle of 'advance informed agreement' to the transnational transfer of living modified organisms (LMOs).[149] Advance informed agreement must be based on risk assessment, which must be 'carried out in a scientifically sound manner . . . and taking into account recognized risk assessment techniques'.[150] Article 26, however, moves beyond a purely scientific approach to decision making, providing that Parties can take into account (albeit 'consistent with their international obligations') 'socio-economic considerations arising from the impact of living modified organisms on the conservation and sustainable use of biological diversity, especially with regard to the value of biological diversity to indigenous and local communities'. This expands the basis for a decision only in fairly narrow circumstances, but potentially meaningfully, and is backed up by support for the involvement of the public in decision making.[151] Similarly, the Cartagena Protocol's approach to the precautionary principle, whilst very similar to Article 5.7 SPS Agreement, contains no suggestion of provisionality, and refers to the 'uncertainty' as well as 'insufficiency' of scientific knowledge.[152] But the Cartagena Protocol also refers explicitly to the Rio Declaration approach to the precautionary principle, which may imply the limitations of 'serious and irreversible damage' and cost-effectiveness.

---

148 There is a large literature on the Cartagena Protocol. See especially Ruth Mackenzie et al., *An Explanatory Guide to the Cartagena Protocol on Biosafety* (IUCN Environmental Policy and Law Paper No. 46, 2003).

149 Article 7. '"Living modified organism" means any living organism that possesses a novel combination of genetic material obtained through the use of modern biotechnology'; '"Living organism" means any biological entity capable of transferring or replicating genetic material . . .'. The definition of 'modern biotechnology' specifies methods 'that overcome natural physiological reproductive or recombination barriers and that are not techniques used in traditional breeding and selection', Article 3(g), (h) and (i).

150 Articles 10 and 15.

151 Article 23(2).

152 'Lack of scientific certainty due to insufficient relevant scientific information and knowledge regarding the extent of the potential adverse effects of a living modified organism on the conservation and sustainable use of biological diversity in the Party of import, taking also into account risks to human health, shall not prevent that Party from taking a decision, as appropriate, with regard to the import of the living modified organism in question . . . in order to avoid or minimize such potential adverse effects', Article 10.6. See also Article 11 for GMOs not intended for intentional introduction into the environment.

Although the precautionary principle is so difficult to pin down that slight differences in wording are unlikely to be too significant, on its face the Cartagena Protocol seems to profess less faith in the ultimate ability of science to resolve uncertainty than the Appellate Body.

LMOs intended for direct use as food or feed or for processing (rather than for 'intentional introduction into the environment') are not subject to advance informed agreement under the Protocol, although decisions can be taken under a 'domestic regulatory framework that is consistent with the objective of this Protocol'.[153] The principle of advance informed agreement is highly significant for trade, but applies therefore to a narrower range of GMOs than the EU regime. And, because it is not primarily concerned with consumer goods, labelling is very much a subsidiary issue in the Cartagena Protocol. Shipments of LMOs destined for direct use as food or feed or for processing are subject to the very limited information requirement that they be labelled 'may contain' LMOs; those destined for contained use or intentional introduction into the environment are simply identified as LMOs.[154]

The difference in emphasis between the Cartagena Protocol and the SPS Agreement may suggest some space for regulatory discretion and an expanded perspective on how GMOs affect social values. The most important contribution of the Cartagena Protocol to the EU's position, however, is the international acknowledgment that case-by-case authorisation on the basis of a range of criteria is an appropriate approach to GMOs. But the place of the Cartagena Protocol in a trade dispute has always been controversial. The preamble is ambiguous, stating that the Protocol does not imply 'a change in the rights and obligations of a Party under any existing international agreements' but also that this comment 'is not intended to subordinate this Protocol to other international agreements'. In *EC–Biotech* the EC argued 'that the Protocol's provisions on precaution and risk assessment inform the meaning and effect of the

---

153 Provision is made for risk assessment and the operation of the precautionary principle, and information on decisions must be provided to a 'Bio-Safety Clearing House'. The objective of the Protocol can be found in Article 1: 'In accordance with the precautionary approach contained in Principle 15 of the Rio Declaration on Environment and Development, the objective of this Protocol is to contribute to ensuring an adequate level of protection in the field of the safe transfer, handling and use of living modified organisms resulting from modern biotechnology that may have adverse effects on the conservation and sustainable use of biological diversity, taking also into account risks to human health, and specifically focusing on transboundary movements'.

154 Article 18. Decision MOP BS-III/10, *Handling, transport, packaging and identification of living modified organisms*: paragraph 2 (a) of Article 18 'requests' parties to use information such as commercial invoices or documentation required in domestic systems and sets out more detailed identification conditions.

relevant provisions of the WTO agreements',[155] emphasising the potential for the Cartagena Protocol to influence interpretation. The Panel, however, concluded that it was not obliged to take into account rules of international law which 'are not applicable to one of the parties to the dispute'.[156] In fact the Panel went further, taking the position that it is obliged to take other international agreements into account only if all WTO members are party to the international agreement.[157] But, although not *bound* to consider the Protocol, 'the mere fact that one or more disputing parties are not parties to a convention does not necessarily mean that a convention cannot shed light on the meaning and scope of a treaty term to be interpreted'.[158] The Panel thereby left itself free to consider international rules 'if it deems such rules to be informative'.[159] This goes some way to rationalising the Panel's approach with the use by the Appellate Body in *Shrimp/Turtle* of international conventions not ratified by all parties to the dispute (let alone all members of the WTO). It also vastly increases the Panel's freedom of action. And with no explanation, the Panel declined to exercise its discretion to consider the Cartagena Protocol: 'Ultimately, however, we did not find it necessary or appropriate to rely on these particular provisions in interpreting the WTO agreements at issue in this dispute'.[160] The use of international treaties by the Appellate Body in *Shrimp/Turtle* brought external values into the trading system, environmental values in that particular dispute. After *Gambling* and *EC–Biotech* it is not clear how easy that will be. This lack of clarity in trade law, with its powerful enforcement mechanisms, makes the collective pursuit of collective goods through multilateral negotiation much more difficult.

The Cartagena Protocol is not, however, the only relevant international material on GMOs. *Shrimp/Turtle* hints at the significance of international negotiation for the purposes of exception under Article XX. Further impetus to negotiation and agreement can be found in the special status of international standards under the SPS and TBT Agreements. Article 3.1 SPS provides that, for the purpose of harmonising SPS measures 'on as wide a basis as possible', Members 'shall base' their SPS measures on international standards, guidelines or recommendations, where they exist, and domestic measures that

---

155 *EC–Biotech*, above n. 4, para. 7.55. See Margaret A. Young, 'The WTO's Use of Relevant Rules of International Law: An Analysis of the *Biotech* Case' (2007) *International and Comparative Law Quarterly*, forthcoming, for a far more detailed and very useful analysis of the Panel's approach to international law.

156 *EC–Biotech*, above n. 4, para. 7.71.

157 *EC–Biotech*, above n. 4, para. 7.68.

158 *EC–Biotech*, above n. 4, para. 7.94.

159 *EC–Biotech*, above n. 4, para. 7.93.

160 *EC–Biotech*, above n. 4, para. 7.95.

'conform with' international standards are rebuttably presumed to be lawful under Article 3.1. Article 3.3 permits members to apply higher standards 'if there is a scientific justification, or as a consequence of the level of sanitary or phytosanitary protection a Member determines to be appropriate'. The TBT Agreement similarly emphasises the desirability of international standards, requiring first that, when 'relevant international standards exist or their completion is imminent', these international standards should be used 'as a basis' for domestic technical regulations, unless they would be 'an ineffective or inappropriate means for the fulfillment of the legitimate objectives pursued, for instance because of fundamental climatic or geographical factors or fundamental technological problems'.[161] When a technical regulation is aiming at one of the 'legitimate objectives' *explicitly* listed in the TBT Agreement, and is 'in *accordance* with relevant international standards', there is a rebuttable presumption that there is no 'unnecessary obstacle to international trade'.[162] And, finally, international standards determine the obligations of notification and consultation in the SPS and TBT Agreements, which are invoked when the content of a proposed SPS measure 'is not substantially the same' as an international standard, or the technical content of a proposed technical regulation is not 'in accordance with' the international standard.[163]

These incentives to use international standards could endow those standards with real normative force. The SPS Agreement refers particularly to standards set by three bodies, the Codex Alimentarius Commission, the International Office of Epizootics, and bodies within the framework of the International Plant Protection Convention,[164] as well as standards 'promulgated by other relevant international organizations open for membership to all Members'.[165] No such list is provided in the TBT Agreement, although an 'international body' must be open to participation by all WTO members.[166]

There are potentially many different bodies for the setting of standards. The Codex Alimentarius Commission (Codex) provides a useful study of the problems and potential of the ambiguous status of these bodies, and is relatively

---

161 Article 2.4. 'The question of effectiveness bears upon the *results* of the means employed, whereas the question of appropriateness relates more to the *nature* of the means employed', *Sardines*, above n. 49, para. 285.

162 Article 2.5, emphasis added.

163 SPS Agreement, Annex B(5). Note also Article 5.8, which provides for the request of an explanation of measures that are not 'based on' international standards. TBT Agreement, Article 2.9.

164 See David G. Victor, 'The Sanitary and Phytosanitary Agreement of the World Trade Organization: An Assessment after Five Years' (1999–2000) 32 *New York University Journal of International Law and Politics* 865, for discussion.

165 Annex A.

166 Annex 1.

active in the debate over GMOs. The Codex deals with a narrow range of public goods, primarily food safety, as well as consumer protection more generally in respect of food. The Codex potentially enhances the status of these values in the WTO system. We know, though, that action in one policy area can have myriad effects on other issues, environmental, economic, social and distributional. The hiving-off of food safety is perfectly sensible if the Codex provides a wholly voluntary standard for consideration by domestic regulators as one factor alongside other public objectives, but becomes less so if that standard then has special status. The Codex also enhances the scientific focus of the WTO, since it perceives and presents itself as a primarily science-based regime:

> The food standards, guidelines and other recommendations of Codex Alimentarius shall be based on the principle of sound scientific analysis and evidence . . .[167]

Mainly at the urging of the EU, however, the relevance of 'other legitimate factors' has been brought into the Codex system:

> When elaborating and deciding upon food standards Codex Alimentarius will have regard, where appropriate, to other legitimate factors relevant for the health protection of consumers and for the promotion of fair practices in food trade.[168]

The Codex demands an explanation of 'how these factors affect the selection of risk management options and the development of standards, guidelines and related texts'. And, whilst 'it should be recognized that some legitimate concerns of governments when establishing their national legislation are not generally applicable or relevant world-wide', only factors accepted 'on a world-wide basis' should make it into the Codex. The acceptance of other legitimate factors is anyway somewhat ambivalent: other legitimate factors should not 'create unjustified barriers to trade', nor should they 'affect the scientific basis of risk analysis'. The scope of other legitimate factors is, moreover, apparently very narrow. They should be 'relevant for the *health* protection of consumers and for the promotion of fair practices in food trade'. The Codex is concerned with food safety and trade, not with environmental protection, sustainable development, or any other social goods, and the notion of fair practices is apparently focused on consumers rather than producers. The role of other legitimate factors remains ambiguous, and although its presence in the

---

[167] Statements of Principle Concerning the Role of Science in the Codex Decision-Making Process and the Extent to which Other Factors are Taken into Account, found in Codex Alimentarius Commission, *16th Procedural Manual*. A 17th version is forthcoming.

[168] Ibid.

international arena is potentially helpful when the EU seeks to justify its regulation of GMOs, the Codex approach is not particularly expansive.

Concerns about the legitimacy of Codex decisions mean that the status of standards in the SPS and TBT Agreements has to be a sensitive issue. Whilst it may have had a useful governance function in the exchange of information and in international collaboration, the Codex was largely neglected by lawyers, scholars and indeed the public until recently. More interest followed the emphasis on international standards in the SPS and TBT Agreements, compounded perhaps by the discussion of controversial issues like GMOs, which increased the visibility of politics in what might once have been perceived as a purely technical body. Although Codex standards are usually adopted by consensus, they can be adopted by majority vote, and sometimes are, including for example the enormously controversial hormones standards. The Appellate Body has explicitly confirmed that 'standards' includes standards adopted by majority.[169] The Codex has been an important forum for discussion of GMOs, with high stakes for, especially, the EU and the US.[170] The sort of deliberation and debate we see in the Codex can lead to common ground and collaborative governance of difficult issues. The nature of the disagreement on GMOs, however, means that consensus is unlikely. But it is possible to envisage the EU or the US pushing something through by majority, and applying such a contentious standard through the WTO dispute settlement machinery would obviously be controversial. There are other process concerns about the Codex. The legitimacy and accountability of the Codex to the interests it affects is increasingly closely scrutinised – and generally found wanting.[171] The main criticisms of the Codex include under-representation of both developing countries and consumer, environmental and other 'outsider' perspectives, as well as over-representation of industry. Even when the relatively excluded groups do attend, lack of resources is likely to limit their effectiveness.

Granting special status to standards is especially strange when considered alongside the irrelevance of the Cartagena Protocol, which for all its faults was negotiated in a context of almost frenzied openness. But the application of any international rules (Cartagena Protocol or standards) in the absence of national

169 *Sardines*, above n. 49; *Hormones*, above n. 54.

170 See Ad Hoc Intergovernmental Task Force on Food Derived from Biotechnology, *Report of Session 7* (2007), available on the Codex website, http://www.codexalimentarius.net/web/archives.jsp?lang=en (accessed December 2007).

171 See the discussion of negative and positive characteristics of the Codex by Michael A. Livermore, 'Authority and Legitimacy in Global Governance: Deliberation, Institutional Differentiation and the Codex Alimentarius' (2006) 81 *New York University Law Review* 766.

consent raises serious questions of sovereignty. The Appellate Body has so far been cautious about hardening the implications of international standards.[172] First, overturning the Panel's interpretation of this part of the SPS Agreement, the Appellate Body in *Hormones* took a flexible approach to 'based on', which certainly does not mean 'conform with'.[173] And in *Sardines*, the Appellate Body held that standards can be a 'basis' for regulation, a starting point, without also being the substantive end point of national regulations, although 'there must be a very strong and very close relationship between two things in order to be able to say that one is "the basis for" the other'.[174] The Appellate Body has also, and more significantly, confirmed that neither Article 3 SPS nor Article 2.4 TBT endows international standards with mandatory force.[175] Hence, a complainant has to establish a breach of the Agreement whatever the challenged measure's relationship with international standards; in the case of the SPS Agreement this means establishing failure to comply with the risk assessment provisions. The Appellate Body has so far refused to interpret the SPS or the TBT Agreement's preference for international standardisation as demanding a reversal of the burden of proof in respect of measures that fail to comply.

But the precise status of standards is not finally settled, and the Appellate Body has confirmed the 'very important role' of international standards in facilitating international trade.[176] It is still possible that the Appellate Body might further enhance the status of international standards. If so, it should also police more effectively the process and legitimacy of standard-setting bodies, making access into the WTO regime conditional on key elements of good governance,[177] such as transparency, revisability and participation, as well as scientific excellence. And just because a standard is not directly applied or policed in WTO disputes does not mean that it is not highly influential. Margaret Young's analysis of *EC–Biotech* makes it clear that the Panel relies

---

172 Joanne Scott, 'International Trade and Environmental Governance: Relating Rules (and Standards) in the EU and the WTO' (2004) 15 *European Journal of International Law* 307.

173 *Hormones,* above n. 54.

174 *Sardines*, above n. 49, para. 245. The Appellate Body rejects the EC argument that only a 'rational relationship' is required. In this case, the Appellate Body was able to avoid determining how close a connection (substantively) there has to be between the international and national standards (because it found that the EC approach actually contradicted the international standards).

175 *Hormones,* above n. 54, para. 165, and *Sardines*, above n. 49, para. 275, respectively.

176 *Sardines*, above n. 49, para. 215.

177 See Scott, above n. 172. Livermore, above n. 171, also argues that the WTO bodies can enhance legitimacy by procedural protection.

heavily on a range of international sources (including the Codex and other standard-setting bodies) and dictionary definitions in its interpretation of the SPS Agreement. Its unconstrained selection of sources can easily lead to 'decontextualised and arbitrary reasoning'.[178]

The role of international law and international standards in the WTO is getting more complicated, not less. And less, not more, transparent – it is important to look at what decision makers are actually doing as well as what they say they are doing. The role of international agreements and standards is not a trivial difficulty, but reminds us that trade is intrinsically linked with the pursuit of other social goods and cannot be addressed separately from them.

BEYOND THE PANEL RULING: THE 'NEW' LEGISLATION

The EU was in a weak position in the *EC–Biotech* dispute: there was a complete absence of transparency and accountability in the moratorium, the niceties of lawful and even-handed rule application had been abandoned, and the safeguard measures were being challenged in EU as well as WTO law. But some powerful and important questions were raised, and the Panel's stance was disappointing in a number of respects. This is not likely to be the end of WTO involvement in GMOs. The new set of laws is also controversial with trading partners, and is also likely to be challenged.[179] It is however a regime that is far more defensible in its own right.

As discussed above (pp. 189–90), it is not obvious which parts of which agreement will apply to the new regime. The SPS Agreement is likely to dominate, not least because the *EC–Biotech* Panel interpreted 'SPS measure',[180] and hence the application of the Agreement, extremely broadly.[181] Most importantly for current purposes, the Panel found that an environmental objective does not preclude application of the SPS Agreement. So an environmental measure applied to protect plants and animals (including, for example, microbes in soil) might be a SPS measure,[182] and the 'residual' category of

178 Young, above n. 155. See also above n. 125.

179 See for example the concerns raised in G/SPS/GEN/204/Rev.7/Add.2. Shaffer, above n. 135, discusses the Panel's strategies for avoiding making a substantive decision.

180 The definition is set out above, n. 46.

181 See Peel, above n. 82, for a nice discussion and contextualisation of the breadth of the Panel's approach. Young, above n. 155, explores in detail the role of international organisations and dictionary definitions in this expansive interpretation.

182 *EC–Biotech*, above n. 4, para. 7.207.

'other damage' (from risks associated with 'pests' specifically) extends this to environmental damage that does not involve damage to the life or health of living organisms, for example damage to geochemical cycles or the dynamics of populations of species.[183] A food risk is also interpreted by the Panel to include what we might more obviously describe as an environmental risk, that is, the risk associated with insects or wild animals consuming GM plants or pollen.[184] And, beyond the place of environmental protection measures in the SPS Agreement, 'pest' is broadly defined to include a plant growing where it is undesired, and hence measures addressing competitive advantage, persistence or invasiveness of GM plants are covered.[185] The product subject to control need not itself be the 'pest', which could apply the SPS Agreement to concerns about a 'crossbreed' enjoying competitive advantages because of genetic modification[186] or increased pesticide resistance in certain insects.[187] The residual category of 'other damage' includes property or economic damage, and so applies the SPS Agreement to the economic impact of coexistence problems, especially given the broad definition of 'pest'.[188]

This 'seismic shift'[189] in our understanding of the reach of the SPS Agreement has obvious implications for the likelihood that the EU regulation of GMOs will be assessed primarily under the SPS Agreement. The very requirement for authorisation, with its implicit ban on marketing pending the successful completion of a risk assessment, as well as refusals or conditional authorisations of individual GMOs, are fundamental questions not considered by the Panel. They are both likely to be addressed under (at least) the SPS Agreement. The EU legislation is highly dependent on scientific disciplines, and the intention of the EU seems to be to comply with, rather than to stretch or challenge, SPS risk assessment disciplines. The answer in individual cases will depend largely on the approach to the science in individual cases. However, in an indication of the depth of the disagreement and distrust here, the US objects even to the *possibility* of the Commission disagreeing with EFSA's opinion on an application, on the basis that this leaves 'room for political interference of the types that had led to the existing moratorium on the approval of biotech products'.[190]

---

183 *EC–Biotech*, above n. 4, para. 7.372–5.
184 *EC–Biotech*, above n. 4, para. 7.292.
185 *EC–Biotech*, above n. 4, para. 7.247.
186 *EC–Biotech*, above n. 4, para. 7.255–8.
187 *EC–Biotech*, above n. 4, para. 7.265–7.
188 *EC–Biotech*, above n. 4, para. 7.370, applied in para. 7.2576, in respect of the Austrian safeguard measures.
189 Peel, above n. 82, p. 1025.
190 G/SPS/GEN/204/Rev.7/Add.2, para. 157.

But, if only the SPS Agreement is applied to the regulation of GMOs, the range of public values pursued is reduced, and the (scientific) constraints on their pursuit are increased. In this, the 'SPS imperialism'[191] of *EC–Biotech* is entirely consistent with the growing imperialism of risk in regulation generally over recent decades. The EU legislation (discussed in detail in Chapters 3 and 4) pursues multiple objectives. Harmonisation underpins the legislation, but beyond that the objectives of the Deliberate Release Directive are stated to be 'to protect human health and the environment',[192] expanded under the Food and Feed Regulation to the provision of 'a high level of protection' for 'human life and health, animal health and welfare, environment and consumer interests'.[193] The objective of traceability is 'facilitating accurate labelling, monitoring the effects on the environment and, where appropriate, on health, and the implementation of the appropriate risk management measures including, if necessary, withdrawal of products'.[194] Animal welfare, consumer interests, and probably some aspects of environmental protection take us beyond even a very extensive interpretation of the SPS Agreement. The EU regime for the control of GMOs pursues a number of, sometimes blurred, objectives. Even the authorisation process for GMOs potentially goes beyond health and safety objectives, and which of the WTO agreements will be used to assess the rules on labelling and traceability is especially difficult. To the extent that labelling is linked to the purpose of protecting human, plant and animal health, it is an SPS measure. Somewhat counter-intuitively (as the labelling rules apply only if the GMO has successfully come through a risk assessment), the Panel held that the Deliberate Release Directive's provisions on labelling are designed to deal with new information or accidental release and so fall within the SPS Agreement. It did not pursue the question of whether labelling under the Deliberate Release Directive also pursued other autonomous objectives, although it found that the efforts to avoid the misleading of consumers under the Novel Food Regulation (which has now been replaced in respect of GM food) applies independently of risk, and so does not fall within the SPS Agreement.

[191] Scott, above n. 8, p. 17.

[192] Directive 2001/18 on the Deliberate Release into the Environment of Genetically Modified Organisms and Repealing Council Directive 90/220/EEC [2001] OJ L 106/1, Article 1.

[193] Regulation 1829/2003, above n. 87, Article 1.

[194] Regulation 1830/2003 concerning the Traceability and Labelling of Genetically Modified Organisms and the Traceability of Food and Feed Products Produced from Genetically Modified Organisms and amending Directive 2001/18/EC [2003] OJ L 268/24, Article 1.

We are bound to turn to the GATT and the TBT Agreement in search of a home for the full breadth of objectives pursued by EU regulation of GMOs.[195] The non-discrimination framework of both of these Agreements brings into play the difficult question, not considered in *EC–Biotech,*[196] of whether GM products are 'like' their conventional counterparts. Less favourable treatment of 'like' products is *prima facie* not permissible, whilst if products are not alike one would not necessarily expect them to be treated in similar ways. Even whether GM maize differs from conventional maize will be controversial. The application of the EU legislation to products produced *from* GMOs is particularly difficult, as there need be no GM material in the actual product to which the legislation is applied, and there may be no way physically to distinguish between products. This question is based on fundamental disagreement as to the nature of agricultural biotechnology: the EU applies special rules to cornflakes made from GM corn because it considers them profoundly different from other cornflakes; to the US, they are just cornflakes. Not only does the decision one way or the other have enormous implications for the application of WTO law, but it divides those who see agricultural biotechnology as a natural extension of conventional practices from those who see it as 'a major watershed in human intervention in nature'.[197]

There is no bright line of 'likeness', and it is perfectly possible that the answer could vary, or at least be more or less easy to reach, depending on the nature of the GMO. So a chimera might more obviously be different from the creatures it is based on, or a modification of corn with the gene from another vegetable might be treated differently from the modification of corn with a bacterium or fish gene. This, however, is not only complex to apply, but also might imply that the EU approach to regulating the technology *as a technology*, albeit on a case-by-case basis, is wrong. The difficult normative implications of likeness are hard to avoid.

Whether products are alike has not yet been considered under the TBT Agreement, but is much debated under the GATT.[198] The Appellate Body

---

195 The EC notified the labelling and traceability provisions under both the TBT Agreement (which it considered to be the prime agreement) and, in response to requests from other members, the SPS Agreement; see European Commission, above n. 144.

196 The question of discrimination did come up under Argentina's Annex C1(a) second clause claim, but the Panel did not address it (see the explicit statement to that effect, para. 8.3). Scott, above n. 8, suggests that it can be surmised that the Panel implicitly places biotech food in one category and novel non-biotech foods in another, p. 229. If this is some indication of the future approach to GMOs, it is obviously significant.

197 The quotation comes from Agriculture and Environment Biotechnology Commission, *Crops on Trial* (AEBC, 2001), para. 80.

198 The approach to likeness may not be the same under the TBT Agreement and

Report in *Asbestos* provides guidance on assessing whether one product is 'like' another. An 'individual, discretionary judgement' is made case by case, and although no evidence 'should be excluded *a priori* from a panel's examination of "likeness"', four general criteria apply:

(i) the properties, nature and quality of the products;
(ii) the end uses of the products;
(iii) consumers' tastes and habits;
(iv) the tariff classification of the products.[199]

It is fairly clear that different regulatory treatment is given to GM and non-GM products with the same end use. The ease with which differences in 'properties, nature and quality' might be identified will vary from case to case, although in some cases of heavily processed products there will be no way of physically distinguishing between regulated (GM) and unregulated (non-GM) products.

The role of consumer tastes and habits is most interesting. One difficult question has always been relevance of process and production methods (PPMs) to the assessment of 'like' products. If the way goods are processed or produced has an impact on the product (for example, harmful pesticide residues in vegetables), there is not likely to be a problem because the 'properties, nature and quality' of the products differ; the difficulty is when there is no obvious impact on the finished product itself. The seminal *Tuna/Dolphin* decision found that tuna fished in a 'dolphin friendly' way is 'like' tuna harvested by methods involving large numbers of dolphin deaths, and hence restrictions on the latter constitute different treatment for 'like products'.[200] This approach discounts those values reflected in the way a product is made, rather than in its physical characteristics once manufactured, and has been heavily criticised.[201] The emphasis of the Appellate Body in *Asbestos* on consumer 'tastes and habits', whilst not addressing PPMs directly, suggests

---

the GATT: famously, likeness 'evokes the image of an accordian', squeezed more or less tightly according to context, *Japan – Taxes on Alcoholic Beverages* (WT/DS10/AB/R, WT/DS11/AB/R, WT/DS8/AB/R, 4 October 1996), in respect of the difference between Article 3.2 and 3.4.

199 *Asbestos*, above n. 6, para. 101.

200 *Tuna/Dolphin*, above n. 13. These reports were never adopted by the GATT membership. Before 1994 and the WTO Dispute Settlement Understanding, Panel rulings had to be adopted before they became binding; adoption now is automatic unless the WTO membership votes unanimously to block the adoption.

201 Robert Howse and Donald Regan, 'The Product/Process Distinction – An Illusory Basis for Disciplining "Unilateralism" in Trade Policy' (2000) 11 *European Journal of International Law* 249, argue that the approach to PPMs rests on an incorrect interpretation of Article III GATT. By contrast, Sanford E. Gaines, 'Processes and

that PPMs affecting consumer decisions *may* be a legitimate basis for distinguishing between products that would otherwise be alike. Consumer preferences in respect of asbestos revolved around the health risks associated with asbestos.[202] It is not far-fetched to imagine the application of this approach to GM crops and food: consumers treat GM and non-GM products differently because (in part) of possible health impacts. The health risk posed by asbestos, however, is notorious and uncontroversial, whilst the health risk posed by GMOs is highly contentious. The accuracy of public or consumer perceptions of risk is likely to be challenged, and this raises the whole question of responses to consumer views that are *not* based on established 'risk'. Because, in any event, consumer views of GMOs may rest on broader public values and on complex uncertainties, taking us into rather different territory from *Asbestos*. But the Appellate Body's concern with 'consumer tastes and habits' is not necessarily limited by the reasons for those tastes and habits.

Proponents of biotechnology are likely to argue that regulation encourages and confirms consumer differentiation. And, as in the EU (p. 85), there is a consciousness in the WTO that this should not be permitted. This concern is captured by the Panel in *Sardines*, in respect not of likeness but of whether rules are necessary at all to prevent the misleading of consumers:

> the danger is that Members, by shaping consumer expectations through regulatory intervention in the market, would be able to justify thereafter the legitimacy of that very same regulatory intervention on the basis of the governmentally created consumer expectations.[203]

In *Sardines*, the EC had failed to produce evidence that 'consumers in most member States have always associated the common name 'sardines' exclusively with *sardina pilchardus*'.[204] There are obvious unanswered questions here, as to the proportion of consumers (or the public) that have to have concerns, the intensity of those concerns, the distribution of those concerns (how many Member States?), but clearly the Appellate Body is willing to scru-

---

Production Methods: How to Produce Sound Policy for Environmental PPM-based Trade Measures' (2002) 27 *Columbia Journal of Environmental Law* 383, argues that the proper focus should be on Article XX.

202 The Appellate Body hints that health risks which do not affect consumer tastes and habits may not enter into the consideration of 'like' products, para. 122.

203 Cited by the Appellate Body, above n. 49, para. 305.

204 *Sardines*, above n. 49, para. 290. *Sardines* involved an EC regulation basically limiting the label 'sardines' to a particular species (*sardina pilchardus (Walbaum)*) of fish. This stopped Peruvian exports of *sardinops sagax sagax* to the EU as 'sardines'. As interpreted by the Appellate Body, the 'relevant' Codex standard allows certain species *other* than *sardina pilchardus* to be marketed as sardines with a qualifier such as 'Peruvian' – 'Peruvian sardines'.

tinise the purpose of regulation. Establishing that EU consumers do wish to discriminate between GM and non-GM food should be relatively straightforward, using evidence from opinion polls and deliberative exercises, and consumer and environmental group campaigns, over a number of years.

Looking specifically at labelling raises some very difficult issues. If the EU cannot make the case here, it is difficult to imagine any situation in which labelling will be available other than to address SPS concerns. But, as discussed in Chapter 4, the rules on labelling and traceability are fiercely resisted by sections of the agricultural biotechnology industry, which argue that consumers have no legitimate basis for distinguishing between GM and non-GM products. The US criticised EC proposals in the SPS Committee on the basis that they 'failed to distinguish the protection of health and the environment from perceived consumer desires'.[205] The concern has to be that the turn to science will determine not only the content of SPS regulation but also what the public can legitimately demand to *know*. In Chapter 4 I argue that labelling is an imperfect effort to regulate complex social goods. Nevertheless, in the heavily marketised society in which we find ourselves, this imperfect opportunity to express political, collective sentiments is crucial. Although there will inevitably be some limitations to labelling rules, if only to avoid absurd requirements, ultimately, if a democracy-sensitive approach to national regulation banning the import of a product is appropriate, that is even more so with labelling. In the case of GMOs, labelling is anyway a 'last resort' response to extreme antipathy to the new technology. As discussed in Chapter 4, consumer choice through labelling underpins most efforts by the EU to justify to its own publics the commercialisation of GMOs, and is used to sweep up all of the public concerns that are difficult to address in the authorisation process.

CONCLUSIONS

The depth and scope of disagreement over GMOs makes this topic an unwelcome responsibility for the WTO. It implicates questions of democracy and the role of government in the WTO system, as well as conflicting and sometimes hidden views on technological development and corporate power. But, if Sheila Jasanoff is correct, regulatory difference is based on profound cultural and political distinctions, and is not going to be wished away.[206] The

205 G/SPS/GEN/204/Rev.7/Add.2, para. 156.

206 Sheila Jasanoff, *Designs on Nature: Science and Democracy in Europe and the United States* (Princeton University Press, 2005).

WTO legal framework continues to evolve,[207] and its application to the EC's completed regulatory framework is far from self-evident, riven with choice and value judgment. Moreover, the *Hormones* aftermath reminds us that, in cases of high politics, the option of facing retaliatory measures rather than complying with an unwelcome WTO judgment is not to be ruled out. At best this creates a worrying message about the competence of international trade bodies; at worst it instigates a potentially destructive cycle of retaliation. The decision in the second *Hormones* dispute, in which the EC challenges the continued application of retaliatory measures by the US and Canada, seems to have been imminent for years now,[208] and the delay reminds us of the extreme sensitivity of the dispute, even as the fall-out continues. GMOs are no easier as a policy area. Toleration and management of difference seems more likely to be productive than condemnation. If the EU, with its enormous economic power, cannot respond to its citizens on this enormously complex question, those with softer voices have few options.

In their exercise of quasi-judicial powers of review, WTO Panels and the Appellate Body face the legitimacy dilemmas familiar in any judicial review of democratically determined regulation or legislation, in terms of both democratic accountability and the second-guessing of regulation that has been carefully constructed by probably more expert and more inclusive bodies.[209] These dilemmas are writ large in the WTO, physically removed from the domestic constituency's democratic influences and regulatory sensitivities. The WTO framework is by no means a straightforwardly anti-regulation, anti-collective set of rules, although it forces members to justify their culturally and socially specific pursuit of the good life in the context of an economic, market-driven vision of the good life. The end of the neglect of trade law by national media and publics means that the WTO must explain itself to constituencies beyond trade. The burden of persuasion is very high, and conspicuously not met by the Panel in *EC–Biotech*. Even accepting that the 'singularly informal and casual' approach of the US in *Shrimp/Turtle* has nothing on the EU's moratorium,[210] the allegation that the complainants (and by implication the Panel) 'wilfully ignore the social controversies' that led to the EU's difficulties is a fair one.[211]

207 The second Panel decision in *Hormones* is expected shortly, above n. 78, and may well have a lot to say about the issues discussed in this chapter.

208 *Hormones II*, above n. 78. The dates in the procedures can be found at http://www.wto.org/english/tratop_e/dispu_e/cases_e/ds320_e.htm (accessed December 2007).

209 See the discussion in Scott, above n. 62.

210 *Shrimp/Turtle*, above n. 10, para. 181.

211 *EC–Biotech*, above n. 4, para. 4.332.

7. Conclusions

Three main themes have dominated this book. First, the regulation of GMOs needs to capture factors beyond the almost overwhelming risk rubric, raising in turn questions about the relative roles of the public and of experts in regulation. Whilst important insights on the complexity of decision making in areas of high technological complexity have now been absorbed into the mainstream, actually making this meaningful is proving difficult. Secondly, GMOs force us to consider the appropriate level of authority for decision making, in a context of increasingly globalised scrutiny of domestic regulation. And thirdly, the 'regulation' of agricultural biotechnology by no means begins and ends with the procedure for the authorisation of a GMO. Overemphasis of the authorisation process can distract from the distribution of very significant responsibilities, costs and benefits in other areas of law. Realistically, regulation stretches both forwards and backwards from authorisation, and we cannot afford to ignore the legal context in which GMOs exist. These three themes compel the addition of governance questions to the list of possible concerns about GMOs – as well as health, environmental, social, political and ethical concerns, close reflection on the ways in which decisions are taken leaves room for disquiet.

The regulation of GMOs is about far more than science, risk and safety. Safety and environmental protection are enormously important and very complex, but so may be the way a technology distributes risk, benefit and power, globally and locally. And new technologies are almost by definition tied up with profound uncertainties that cannot be resolved by more or better science. This does not mean that we should stop innovating. It does, though, mean that we need to engage with the purposes and objectives of the new technology, overall and in individual cases. The extension of relevant factors for regulation implies in turn that the public as well as experts have a role to play. There are many possible meanings for 'public participation', and many possible purposes, from improving democracy to improving results. But, at the very least, public participation in decision making provides an opportunity to identify, explore and evidence the values at stake in a decision.

Much of this book has been about searching for a forum in which the meaning and implications of GMOs can be given regulatory consideration. The social and ethical issues associated with new technology, including the social and ethical commitments embedded in the pursuit of that new technology, are a necessary part of its regulation. Examining the regulatory framework for GMOs demonstrates that the complexity of decision making and the value basis of decisions are generally recognised in regulatory, policy and legislative circles. The UK's *GM Nation?* debate (Chapter 2), the 'other legitimate factors' formula in the EU legislation (Chapter 3) and the WTO acceptance of values and the demands of democracy (Chapter 6) all respond to decades of social science explaining and evidencing the broad nature of regulatory decision making. The acceptance of the precautionary principle (Chapters 2 and 3) further acknowledges the fragility of scientific information as the sole provider of legitimacy for decisions, and hence implies a more outward-looking process. In the broader context of regulation, coexistence and liability are both capable of addressing certain socioeconomic issues, patents have a 'morality' proviso, and consumer choice is in the background to sweep up any leftover ethical, social or political issues.

We can conclude that the legal management of GMOs is far from wholly technocratic, and is not restricted to the regulation of risk. These are important and exciting legal innovations. But what we should do with these innovations is much more difficult, and still unresolved. The acceptance in principle of the very broad nature of the decision at stake is consistently overwhelmed by the more general legal and political context. When it comes to the crunch, the persistent instinct is to turn to the experts, usually in science, but also experts in economics, and possibly increasingly in ethics. To go back to the examples raised in the previous paragraph, the *GM Nation?* debate received on the whole quite shallow 'evidence-based' responses from the UK Government, albeit within a rhetoric of values and responsiveness. The 'other legitimate factors' formula is subject to legal and political pressures that will make its use extraordinary. And the WTO Panels and the Appellate Body use all the right language, but, when it comes to application in cases of difficult science and difficult politics, take a very narrow approach to the evidence before them. Even the precautionary principle, with its radical potential, is primarily about scientific evidence according to the European Courts.

So the mainstream has now accepted that regulation is legitimately value based, open, participative. But then what actually happens falls some distance short. Sometimes presumably the failure to follow through is cynical. The language of democracy is perhaps manipulated to disguise the narrow basis for decisions that will allow only physical harms to justify economic risks. Whilst, as discussed in Chapter 1, public opinion is accepted as real and something that must be addressed for political reasons, the reality of what lies behind opinion is still not properly grasped. This is not necessarily wholly cynical. Longstanding routines and habits of decision-making institutions have very firmly entrenched the dichotomies of 'real risk' versus 'perceived risk' and 'rational expert' versus 'irrational public'. Tacking the new understanding of the nature of the questions being asked onto old approaches of regulation is bound to be inadequate. The basic premise of that change is accepted, but as if it were a small evolutionary reform, when actually the change implied is revolutionary. The language of values and the language of participation have been thoroughly learned. Their meaning is far more challenging. The focus now has to be on working through the implications of change, and identifying what happens to public concerns articulated in the regulatory arena. This demands engagement also with the implicit values and commitments in the 'science-based' approach to regulation, not least its failure to engage with uncertainty. The importance of economic growth as a form of progress and the contribution of agricultural biotechnology to economic growth also both absolutely saturate policy, but are never critically examined.

Another constant tension observed throughout this book is between different levels of regulation. The democratic pressures are most intense at local or national level, and it is increasingly difficult for 'the public' to be heard as we internationalise decision making. It is this public, though, that most readily captures the non-scientific aspects of decisions and is most alert to the uncertainty that pervades regulation of GMOs. The higher up the international system, and the more removed from 'politics' (or democratic influence) as traditionally conceived, the more attractive a scientific basis for decision making becomes. But to argue that the full breadth of concerns should be addressed by regulation means also that the democratic will should be acknowledged and respected in decision making. This in turn means that a way needs to be found to accommodate national perspectives in a process of international governance. This is extraordinarily difficult within the terms of the trade systems so laboriously built up in recent decades. But our systems of democratic control and accountability have been built up even more slowly, and sometimes painfully. New technologies raise questions about the social purposes of free trade, at regional and global levels. Whilst the surrendering of control over certain issues is accepted for the benefits of membership, the benefits of trade, it should not be used to discipline 'irrational' politics. It is imperative not to be pessimistic about the role of the state or the EU to respond to these outside pressures. The EU has strong objectives that go beyond free trade, and even in the WTO the pursuit of free trade is far from unequivocal. Domestic actors are not the hapless pawns of globalisation, and debate over the content and application of the rules makes for a continuous process of contestation, a constant dynamism and change. The EU, the nation state, subnational and global institutions interact in a complex web of finely determined responsibilities, such that identifying any simple apex of authority is impossible.

Expanding the grounds available for national (or EU in the WTO context) autonomy is the most modest way to accommodate more fully domestic perspectives in the EU and WTO systems. There is space in these systems because, again, the importance of national responsiveness to democratic concerns and the complexity of public responses to technology are largely accepted in principle. But again, this proves very difficult to put into practice. In practice, domestic authorities are usually forced to justify their regulatory measures in a trade context primarily on scientific grounds. This goes to the heart of the transparency and accountability of decision making. Not only is the depth and extent of public concern unlikely to receive a full response from government, but it is possible or even likely that political decisions will be dressed up as scientific decisions. The loss of accountability and contestability should be of as great concern to advocates of a strong and predictable trading system as to advocates of local democratic responsiveness.

Demanding a response to 'public will' is not the same as saying that the 'public' should be followed. The public is difficult to identify, rarely speaks with one voice, and may be manipulated or simply wrong. There is a dark side to public participation in decision making, in this case the danger of missing the social benefits of technological development. And we see increasing reference in EU and national policy to the potential public benefits of agricultural biotechnology, especially feeding the poor and responding to climate change. The purpose of and need for a development should be a key factor in regulation. Proponents of biotechnology cannot, however, expect assessment of need to be uncritical. Assertions of the future social promise of biotechnology, although often presented as straightforward scientific fact, are value laden and uncertain. In any event, there is no necessary link between a large multinational's application for authorisation of herbicide-resistant maize and the still unfulfilled public potential of biotechnology.

Extending the scope of decisions on GMOs and enhancing the domestic role in those decisions go together. These changes are not easy and not without risks. But the opposite risk is fracturing national authority, and hence of the democracy that we currently enjoy. This in turn creates risks for the international trading system. EU and WTO systems enjoy only fragile authority and have no deep reserves of legitimacy to call on in tough times. GMOs are already something of a cipher for public anger, and if they are pushed onto unwilling publics through the highest-profile institutions of globalisation, the harm could be serious.

Regulation is about much more than the process of authorising (or not) a new technology. It must also be about what happens *around* and *after* authorisation. So rules on labelling, coexistence, liability and intellectual property are a crucial part of the regulatory settlement for GMOs, influencing the relationship between the biotechnology industry and those it affects. A satisfactory set of conditions for the commercialisation (or not) of GMOs demands that attention be paid to a number of legal disciplines, and dialogue between these disciplines is vital for the full consideration of the conditions in which GMOs are cultivated and marketed. In fact it is difficult to see how authorisation decisions can be made at all without discussion of a reasonably predictable post-authorisation regime. But they are, all the time, and a predictable post-authorisation regime is some distance away. Looking at other legal disciplines may also help us to move beyond concentration on a narrow set of technical criteria for the assessment of GMOs. But, as things stand, these forms of regulation also fail to grasp properly the implications of decisions. We should be conscious of the lawyers' complicity in the marginalisation and control of broader social debate about GMOs. It is especially in the conditions and trade law implications of coexistence, in liability rules and in patent rules, that deeply political choices are most misrepresented as discrete technical legal questions, suitable for neutral and sheltered application by experts. WTO rules are also obscure and resistant to non-specialist engagement. We have our very own technocracy, and it is important that these areas of regulation are available to social debate. The burden of persuasion is considerable.

Much of this book has been about the search for a forum in which the broad range of concerns about GMOs can be fed into regulation. There are, at best, gaps in the regulatory consideration of what is important about agricultural biotechnology. At worst, the regulation simply ignores or fails to appreciate the social pressures it faces. But it would be churlish not to accept that there are increasing efforts to engage. And it is important to be optimistic. The legal and political framework need not preclude the consideration of the full range of issues provoked by GMOs. It is the context in which the regulation operates that overwhelms politics with risk and underplays the purposes and objectives of GMOs. Social and ethical issues cannot be hived off into a separate and self-contained space. The full debate on GMOs should be capable of consideration in the mainstream, at every step of the regulatory process.

Bibliography

Ad Hoc Intergovernmental Task Force on Food Derived from Biotechnology, *Report of Session 7* (2007) at www.codexalimentarius.net/web/archives.jsp?lang=en.

Advisory Committee on Releases to the Environment (ACRE), *Advice on the Implications of the Farm-Scale Evaluations of Genetically Modified Herbicide Tolerant Crops*, 13 January 2004.

Aerts, Rob J., 'The Industrial Applicability and Utility Requirements for the Patenting of Genomic Inventions: A Comparison Between European and US Law' (2004) 26 *European Intellectual Property Review* 349.

Agriculture and Environment Biotechnology Commission, *Crops on Trial* (AEBC, 2001).

Agriculture and Environment Biotechnology Commission, *What Shapes the Research Agenda? In Agricultural Biotechnology* (AEBC, 2005).

Aplin, Tanya and Jennifer Davis, *Cases and Materials on Intellectual Property* (Oxford University Press, 2008, forthcoming).

Applegate, John S., 'The Prometheus Principle: Using the Precautionary Principle to Harmonize the Regulation of Genetically Modified Organisms' (2001) 9 *Indiana Journal of Global Legal Studies* 207.

Atwood, Margaret, *Oryx and Crake* (Virago, 2003).

Barry, Andrew, *Political Machines: Governing a Technological Society* (The Athlone Press, 2001).

Beck, Ulrich, *Risk Society: Towards a New Modernity* (Sage Publications, 1992).

Beckett, Margaret, Secretary of State, Statement on GM Policy, 9 March 2004.

Bently, Lionel and Brad Sherman, *Intellectual Property Law* (Oxford University Press, 2001).

Berglund, Christer and Simon Matti, 'Citizen and Consumer: the Dual Role of Individuals in Environmental Policy' (2006) 15 *Environmental Politics* 550.

Bernauer, Thomas, *Genes, Trade, and Regulation: The Seeds of Conflict in Food Biotechnology* (Princeton University Press, 2003).

Better Regulation Commission, *Risk, Responsibility and Regulation – Whose Risk is it Anyway?* (2006).

Black, Julia, 'Regulation as Facilitation: Negotiating the Genetic Revolution' (1998) 61 *Modern Law Review* 621.

Blair, Tony, 'Britain's Path to the Future – Lit by the Brilliant Light of Science', 3 November 2006, The King's Centre, Oxford.

Blakeney, Michael, 'The Protection of Traditional Knowledge under Intellectual Property Law' (2000) 22 *European Intellectual Property Review* 251.

Boyle, James, 'The Second Enclosure Movement and the Construction of the Public Domain' (2003) 66 *Law and Contemporary Problems* 33.

Bradley, Kieran St Clair, 'Institutional Aspects of Comitology: Scenes from the Cutting Room Floor' in Christian Joerges and Ellen Vos (eds), *EU Committees: Social Regulation, Law and Politics* (Hart Publishing, 1999).

Bratspies, Rebecca, 'The Illusion of Care: Regulation, Uncertainty, and Genetically Modified Food Crops' (2002) 10 *New York University Environmental Law Journal* 297.

Bratspies, Rebecca, 'Myths of Voluntary Compliance: Lessons from the Starlink Corn Fiasco' (2003) 27 *William and Mary Environmental Law and Policy Review* 591.

Breyer, Stephen, *Breaking the Vicious Circle: Toward Effective Risk Regulation* (Harvard University Press, 1993).

Brosset, Estelle, 'The Prior Authorisation Procedure Adopted for the Deliberate Release into the Environment of Genetically Modified Organisms: The Complexities of Balancing Community and National Competences' (2004) 10 *European Law Journal* 555.

Burrell, Robert and Stephen Hubicki, 'Patent Liability and Genetic Drift' (2005) 7 *Environmental Law Review* 278.

Cabinet Office, *Field Work: Weighing up the Costs and Benefits of GM Crops*, at www.cabinetoffice.gov.uk/strategy/publications.aspx.

Center for Food Safety, *Monsanto vs. U.S. Farmers* (2004).

Chalmers, Damian, '"Food for Thought": Reconciling European Risks and Traditional Ways of Life' (2003) 66 *Modern Law Review* 532.

Chalmers, Damian, 'Risk, Anxiety and the European Mediation of the Politics of Life' (2005) 30 *European Law Review* 649.

Chambers, Graham R., 'The BSE Crisis and the European Parliament' in Christian Joerges and Ellen Vos (eds), *EU Committees: Social Regulation, Law and Politics* (Hart Publishing, 1999).

Christoforou, Theofanis, 'The Precautionary Principle and Democratizing Expertise: A European Legal Perspective' (2003) 30 *Science and Public Policy* 205.

Christoforou, Theofanis, 'The Regulation of Genetically Modified Organisms in the European Union: The Interplay of Science, Law and Politics' (2004) 41 *Common Market Law Review* 637.

Clapp, Jennifer, 'Transnational Corporate Interests and Global Environmental Governance: Negotiating Rules for Agricultural Biotechnology and Chemicals' (2003) 12 *Environmental Politics* 1.

Clapp, Jennifer, 'The Political Economy of Food Aid in an Era of Agricultural Biotechnology' (2005) 11 *Global Governance* 467.

Cohen, Lizabeth, 'Citizens and Consumers in the United States in the Century of Mass Consumption' in Martin Daunton and Matthew Hilton (eds), *The Politics of Consumption: Material Culture and Citizenship in Europe and America* (Berg, 2001).

Committee of Independent Experts, *Allegations Regarding Fraud, Mismanagement and Nepotism in the European Commission* (European Parliament, 1999)

Commoner, Barry, *Science and Survival* (The Viking Press, 1963).

Cordonnier-Segger, Marie-Claire and Markus W. Gehring, 'The WTO and Precaution: Sustainable Development Implications of the WTO Asbestos Dispute' (2003) 15 *Journal of Environmental Law* 289.

Cornish, William R., 'Experimental Use of Patented Inventions in the EC States' (1998) 29 *International Review of Industrial Property and Copyright Law* 735.

Craig, Paul, 'The Fall and Renewal of the Commission: Accountability, Contract and Administrative Organisation' (2000) 6 *European Law Journal* 98.

Daly, Herman E., 'On Wilfred Beckerman's Critique of Sustainable Development' (1995) 4 *Environmental Values* 49.

Daunton, Martin and Matthew Hilton, 'Material Politics: An Introduction' in Martin Daunton and Matthew Hilton (eds), *The Politics of Consumption: Material Culture and Citizenship in Europe and America* (Berg, 2001).

Daunton, Martin and Matthew Hilton (eds), *The Politics of Consumption: Material Culture and Citizenship in Europe and America* (Berg, 2001).

Davies, Gareth, 'Morality Clauses and Decision-Making in Situations of Scientific Uncertainty: The Case of GMOs' (2006) *Hebrew University International Law Research Paper No. 10-06.*

de Búrca, Gráinne and Joanne Scott (eds), *The EU and the WTO: Legal and Constitutional Issues* (Hart Publishing, 2001).

de Búrca, Gráinne and Joanne Scott, 'The Impact of the WTO on EU Decision-making' in Gráinne de Búrca and Joanne Scott (eds), *The EU and the WTO: Legal and Constitutional Issues* (Hart Publishing, 2001).

de Búrca, Gráinne and Joanne Scott (eds), *Law and New Governance in the EU and the US* (Hart Publishing, 2006).

de Sadeleer, Nicholas, 'Procedures for Derogations from the Principle of Approximation of Laws under Article 95 EC' (2003) 40 *Common Market Law Review* 889.

de Sadeleer, Nicholas, 'The Precautionary Principle in EC Health and Environmental Law' (2006) 12 *European Law Journal* 139.

Dehousse, Renaud, 'Regulation by Networks in the European Community:

The Role of European Agencies' (1997) 4 *Journal of European Public Policy* 246.
Dent, Chris, 'Decision-Making and Quality in Patents: An Exploration' (2006) 28 *European Intellectual Property Review* 381.
Department for the Environment, Food and Rural Affairs (DEFRA), *Government Response to the Environmental Audit Committee Report: GM Foods – Evaluating the Farm Scale Trials*, Second Report, Session 2003–04.
Department for the Environment, Food and Rural Affairs (DEFRA), *The GM Dialogue: Government Response*, 9 March 2004.
Department for the Environment, Food and Rural Affairs (DEFRA), *Consultation on Proposals for Managing the Coexistence of GM, Conventional and Organic Crops* (DEFRA, 2006).
Departmènt of Trade and Industry, *Response to the Royal Society and Royal Academy of Engineering Report: 'Nanoscience and Nanotechnologies: Opportunities and Uncertainties'* (DTI, 2005).
Dobson, Andrew, *Citizenship and the Environment* (Oxford University Press, 2003).
Dobson, Andrew, 'Biocentrism and Genetic Engineering' (1995) 4 *Environmental Values* 3.
Drahos, Peter, 'Biotechnology Patents, Markets and Morality' (1999) 21 *European Intellectual Property Review* 441.
Drahos, Peter, *Towards an International Framework for the Protection of Traditional Group Knowledge and Practice* (UNCTAD–Commonwealth Secretariat Workshop on Elements of National Sui Generis Systems for the Preservation, Protection and Promotion of Traditional Knowledge, Innovations and Practices and Options for an International Framework, 2004).
Dutfield, Graham, 'The Public and Private Domains: Intellectual Property Rights in Traditional Knowledge' (2000) 21 *Science Communication* 274.
Dutfield, Graham, *Intellectual Property Rights and the Life Science Industries: A Twentieth Century History* (Ashgate, 2003).
Dutfield, Graham, 'Should We Regulate Technology through the Patent System? The Case of Terminator Technology' in Han Somsen (ed.), *The Regulatory Challenge of Biotechnology: Human Genetics, Food and Patents* (Edward Elgar, 2007).
Echols, Marsha, 'Food Safety Regulation in the European Union and the United States: Different Cultures, Different Laws' (1998) 4 *Columbia Journal of European Law* 525.
Eckersley, Robyn, 'A Green Public Sphere in the WTO: The *Amicus Curiae* Interventions in the Trans-Atlantic Biotech Dispute' (2005) *Ecolomic Policy and Law: Journal of Trade and Environment Studies*, at www.ecolomics-international.org.

Eisenberg, Rebbeca, 'Patents and the Progress of Science: Exclusive Rights and Experimental Use' (1989) 56 *University of Chicago Law Review* 1017.

Environment, Food and Rural Affairs Select Committee, Session 2002-2003 18th Report, *Conduct of the GM Public Debate*, HC 1220.

Environmental Audit Committee, 2003–2004 Second Report, *GM Foods – Evaluating the Farm Scale Trials* HC 90-I.

European Commission, *Biotechnology in the Community* COM (1983) final.

European Commission, *Communication on the Precautionary Principle* COM (2000) 1 final.

European Commission, *Managing Natura 2000 Sites* (European Communities, 2000).

European Commission, *White Paper on Environmental Liability* COM (2000) 66 final.

European Commission, *White Paper on European Governance* COM (2001) 428 final.

European Commission, *Proposal for a Regulation of the European Parliament and of the Council on Genetically Modified Food and Feed* COM (2001) 425 final.

European Commission, *Science and Society Action Plan* (2002).

European Commission, *Life Sciences and Biotechnology – A Strategy for Europe* COM (2002) 27 final.

European Commission, *Communication on the Collection and Use of Expertise by the Commission: Principles and Guidelines 'Improving the Knowledge Base for Better Policies'* COM (2002) 713 final.

European Commission, *Agriculture and Environment* (2003).

European Commission Recommendation, *Guidelines for the Development of National Strategies and Best Practices to Ensure the Co-Existence of Genetically Modified (GM) Crops with Conventional and Organic Farming* C (2003).

European Commission, Staff Working Document, *European Action Plan for Organic Food and Farming* SEC (2004) 739.

European Commission, *European Action Plan for Organic Food and Farming* COM (2004) 415 final.

European Commission, *Working Together for Growth and Jobs: A New Start for the Lisbon Strategy* COM (2005) 24 final.

European Commission, *Proposal for a Directive amending Directive 2001/18 concerning the deliberate release into the environment of genetically modified organisms, as regards the implementing powers conferred to the Commission* COM (2006) 920 final.

European Commission, *Proposal for a Regulation amending Regulation No 1829/2003 on genetically modified food and feed, as regards the implementing powers conferred on the Commission* COM (2006) 912 final.

European Commission, *Report on the Implementation of National Measures on the Coexistence of Genetically Modified Crops with Conventional and Organic Farming* COM (2006) 104 final.

European Commission, *Report on the Implementation of Regulation 1829/2003 on Genetically Modified Food and Feed* COM (2006) 626 final.

European Commission, *Report on the Implementation of Regulation (EC) No. 1830/2003 concerning the Traceability and Labelling of Genetically Modified Organisms and amending Directive 2001/18/EC* COM (2006) 197 final.

European Commission, *Communication on the Mid-term Review of the Strategy of Life Sciences and Biotechnology* COM (2007) 175 final.

European Commission, *Proposal for a Council Decision concerning the placing on the market in accordance with Directive 2001/18/EC of a potato product genetically modified for enhanced content of the amyloprotein component of starch* COM (2007) 336 final.

European Commission, *Second Report on the Experience of Member States with GMOs placed on the Market under Directive 2001/18/EC on the Deliberate Release into the Environment of Genetically Modified Organisms* COM (2007) 81 final.

European Commission, Staff Working Document, *Second Report on the Experience of Member States with GMOs Placed on the Market under Directive 2001/18/EC on the Deliberate Release into the Environment of Genetically Modified Organisms* SEC (2007) 274.

European Commission, Staff Working Document, *Communication on the Mid-term Review of the Strategy on Life Sciences and Biotechnology* SEC (2007) 441.

European Commission, *Mid-term Review of the Strategy on Life Sciences and Biotechnology* COM (2007) 175 final.

European Commission, *Second Report on the Experience of Member States with GMOs Placed on the Market under Directive 2001/18/EC on the Deliberate Release into the Environment of Genetically Modified Organisms* COM (2007) 81 final.

European Food Safety Authority, *Guidance Document of the Scientific Panel on Genetically Modified Organisms for the Risk Assessment of Genetically Modified Plants and Derived Food And Feed* (EFSA, 2006).

European Patent Office, *Guidelines for Examination in the European Patent Office* (2005).

Feintuck, Mike, 'Precautionary Maybe, but What's the Principle? The Precautionary Principle, the Regulation of Risk, and the Public Domain' (2005) *Journal of Law and Society* 371.

Fisher, Elizabeth, 'Precaution, Precaution Everywhere: Developing a "Common Understanding" of the Precautionary Principle in the European

Community' (2002) 9 *Maastricht Journal of International and Comparative Law* 7.

Fisher, Elizabeth, 'Risk and Environmental Law: A Beginner's Guide' in Benjamin Richardson and Stepan Woods (eds), *Environmental Law for Sustainability* (Hart Publishing, 2006).

Fisher, Elizabeth, *Risk: Regulation and Administrative Constitutionalism* (Hart Publishing, 2007).

Footer, Mary, 'A Tale of Two Commons: Plant Genetic Resources and Agricultural Trade Reform' in Han Somsen (ed.), *The Regulatory Challenge of Biotechnology: Human Genetics, Food and Patents* (Edward Elgar, 2007).

Francioni, Francesco and Tullio Scovazzi (eds), *Biotechnology and International Law* (Hart Publishing, 2006).

Friends of the Earth International, *Who Benefits from GM Crops?* (Friends of the Earth International, 2006).

Fukayama, Francis, *Our Posthuman Future: Consequences of the Biotechnology Revolution* (Profile Books, 2002).

Gaines, Sanford E., 'The WTO's Reading of the GATT Article XX Chapeau: A Disguised Restriction on Environmental Measures' (2001) 22 *University of Pennsylvania Journal of International Economic Law* 739.

Gaines, Sanford E., 'Processes and Production Methods: How to Produce Sound Policy for Environmental PPM-Based Trade Measures' (2002) 27 *Columbia Journal of Environmental Law* 383.

Galanter, Marc, 'Why the "Haves" Come Out Ahead: Speculations on the Limits of Legal Change' (1974) 9 *Law and Society* 95.

Glover, Jonathon, *Humanity: A Moral History of the Twentieth Century* (Pimlico, 2001).

GM Nation? The Findings of a Public Debate (AEBC, 2003) at www2.aebc.gov.uk/aebc/reports/gm_nation_report_final.pdf.

GM Science Panel, *GM Science Review: First Report: An Open Review of the Science Relevant to GM Crops and Food Based on the Interests and Concerns of the Public* (2003), at www.gmsciencedebate.org.uk.

GM Science Panel, *GM Science Review: Second Report: An Open Review of the Science Relevant to GM Crops and Food based on Interests and Concerns of the Public* (2004), at www.gmsciencedebate.org.uk.

Gold, E. Richard, 'Biomedical Patents and Ethics: A Canadian Solution' (2000) 45 *McGill Law Journal* 413.

Gold, E. Richard and Alain Gallochat, 'The European Biotech Directive: Past as Prologue' (2001) 7 *European Law Journal* 331.

Gopalakrishnan, N.S., 'TRIPS and Protection of Traditional Knowledge of Genetic Resources: New Challenges to the Patents System' (2005) 27 *European Intellectual Property Review* 11.

Goven, Joanna, 'Processes of Inclusion, Cultures of Calculation, Structures of Power: Scientific Citizenship and the Royal Commission on Genetic Modification' (2006) 31 *Science, Technology and Human Values* 565.

Grove-White, Robin, Philip Macnaghten, Sue Mayer and Brian Wynne, *Uncertain World: Genetically Modified Organisms, Food and Public Attitudes in Britain* (Centre for the Study of Environmental Change & Unilever, 1997).

Grove-White, Robin, Phil MacNaghten and Brian Wynne, *Wising Up: The Public and New Technologies* (Centre for the Study of Environmental Change, Lancaster University, 2000).

Grubb, Philip, *Patents for Chemicals, Pharmaceuticals and Biotechnology: Fundamentals of Global Law, Practice and Strategy* (Oxford University Press, 2004).

Gunningham, Neil, 'Regulating Biotechnology: Lessons from Environmental Policy' in Han Somsen (ed.), *The Regulatory Challenge of Biotechnology: Human Genetics, Food and Patents* (Edward Elgar, 2007).

Habermas, Jurgen, *The Future of Human Nature* (Polity, 2003).

Hamilton, Chris, 'Biodiversity, Biopiracy and Benefits: What Allegations of Biopiracy Tell Us about Intellectual Property' (2006) 3 *Developing World Bioethics* 1471.

Hardin, Garrett, 'The Tragedy of the Commons' (1968) 162 *Science* 1243.

Heath, Christopher and Sabine Weidlich, 'Intellectual Property: Suitable for Protecting Traditional Medicine?' (2003) *Intellectual Property Quarterly* 69.

Heller, Michael A., 'The Tragedy of the Anticommons: Property in Transition from Marx to Markets' (1998) 111 *Harvard Law Review* 621.

Heller, Michael and Rebecca Eisenberg, 'Can Patents Prevent Innovation?' (1998) 280 *Science* 698.

Hervey, Tamara, 'Regulation of Genetically Modified Products in a Multi-Level System of Governance: Science or Citizens?' (2001) 10 *Review of European Community and International Environmental Law* 321.

Heyvaert, Veerle, 'Guidance Without Constraint: Assessing the Impact of the Precautionary Principle on the European Community's Chemicals Policy' (2006) 6 *Yearbook of European Environmental Law* 27.

Heyvaert, Veerle, 'Facing the Consequences of the Precautionary Principle in European Community Law' (2006) 31 *European Law Review* 185.

Hilson, Christopher, 'New Social Movements: The Role of Legal Opportunity' (2002) 9 *Journal of European Public Policy* 238.

Hilson, Christopher, 'Beyond Rationality? Judicial Review and Public Concern in the EU and the WTO' (2005) 56 *Northern Ireland Legal Quarterly* 320.

Hilson, Christopher, 'Information Disclosure and the Regulation of Traded Product Risks' (2005) 17 *Journal of Environmental Law* 305.

Hindmarsh, Richard, Geoffrey Lawrence and Janet Norton, 'Bio-utopia: The Way Forward?' in Hindmarsh, Lawrence and Norton (eds), *Altered Genes; Reconstructing Nature: The Debate* (Allen and Unwin, 1998).

Hindmarsh, Richard, Geoffrey Lawrence and Janet Norton (eds), *Altered Genes; Reconstructing Nature: The Debate* (Allen and Unwin, 1998).

House of Commons Select Committee of Science and Technology, Session 1998–99, 1st Report, *Scientific Advisory System: Genetically Modified Foods.*

House of Lords Select Committee on Science and Technology, Session 2000–01, 3rd Report, *Science and Society.*

Howse, Robert, 'Democracy, Science and Free Trade: Risk Regulation on Trial at the WTO' (2000) 98 *Michigan Law Review* 2329.

Howse, Robert, 'From Politics to Technocracy – and Back Again: The Fate of the Multilateral Trading Regime' (2002) 96 *American Journal of International Law* 94.

Howse, Robert and Petros C. Mavroidis, 'Europe's Evolving Regulatory Strategy for GMOs: The Issue of Consistency with WTO Law: of Kine and Brine' (2000) 24 *Fordham International Law Journal* 317.

Howse, Robert and Donald Regan, 'The Product/Process Distinction – An Illusory Basis for Disciplining "Unilateralism" in Trade Policy' (2000) 11 *European Journal of International Law* 249.

Hughes, Emma, 'Dissolving the Nation: Self-Deception and Symbolic Inversion in the GM Debate' (2007) 16 *Environmental Politics* 318.

Institute for the Study of International Migration, *Genetically Modified Food in the Southern Africa Food Crisis of 2002–2003* (Georgetown University School of Foreign Service, 2004).

Irwin, Alan, 'Constructing the Scientific Citizen: Science and Democracy in the Biosciences' (2001) 10 *Public Understanding of Science* 1.

Irwin, Alan, 'The Politics of Talk: Coming to Terms with the "New" Scientific Governance' (2006) 36 *Social Studies of Science* 299.

Jacoby, Wade and Sophie Meunier, *Europe and the Management of Globalization: Defensive and Offensive Responses to Globalization Pressures*, at www.princeton.edu/~smeunier/conference_europeanization. htm.

Janis, Mark D., 'Sustainable Agriculture, Patent Rights and Plant Innovation' (2001) 9 *Indiana Journal of Global Studies* 91.

Jasanoff, Sheila, *The Fifth Branch: Science Advisers as Policymakers* (Harvard University Press, 1990).

Jasanoff, Sheila, *Designs on Nature: Science and Democracy in Europe and the United States* (Princeton University Press, 2005).

Joerges, Christian, 'Deliberative Political Processes Revisited: What Have We Learnt About the Legitimacy of Supranational Decision-Making?' (2006) 44 *Journal of Common Market Studies* 779.

Joerges, Christian and Renaud Dehousse (eds), *Good Governance in Europe's Integrated Market* (Oxford University Press, 2002).

Joerges, Christian and Jurgen Neyer, 'From Intergovernmental Bargaining to Deliberative Process: The Constitutionalisation of Comitology' (1997) 3 *European Law Journal* 273.

Joerges, Christian and Ellen Vos (eds), *EU Committees: Social Regulation, Law and Politics* (Hart Publishing, 1999).

Joint Research Centre Reference Report, *Consequences, Opportunities and Challenges of Modern Biotechnology for Europe* (JRC European Commission, 2007).

Kearnes, Matthew et al., 'From Bio to Nano: Learning Lessons from the UK Agricultural Biotechnology Controversy' (2006) 15 *Science as Culture* 291.

Kevles, Daniel J., *A History of Patenting Life in the United States with Comparative Attention to Europe and Canada* (European Group on Ethics in Science and New Technologies to the European Commission, 2002).

Kingston, William, 'Why Harmonisation is a Trojan Horse' (2004) 26 *European Intellectual Property Review* 447.

Kirkham, Georgiana, '"Playing God" and "Vexing Nature": A Cultural Perspective' (2006) 15 *Environmental Values* 173.

Krapohl, Sebastian, 'Credible Commitment in Non-Independent Regulatory Agencies: A Comparative Analysis of the European Agencies for Pharmaceuticals and Foodstuffs' (2004) 10 *European Law Journal* 518.

Kritzer, Herbert M. and Susan Silbey, *In Litigation: Do the Haves Still Come Out Ahead?* (Stanford University Press, 2003).

Kysar, Douglas A., 'Preferences for Processes: The Process/Product Distinction and the Regulation of Consumer Choice' (2004) 118 *Harvard Law Review* 525.

Ladeur, Karl-Heinz, 'The Introduction of the Precautionary Principle into EU Law: A Pyrrhic Victory for Environmental and Public Health Law? Decision Making under Conditions of Complexity in Multi-Level Political Systems' (2003) 30 *Common Market Law Review* 1455.

Lamy, Pascal, Director General, *The WTO and its Agenda for Sustainable Development* (Yale University, 24 October 2007) at www.wto.org/english/news_e/sppl_e/sppl79_e.htm.

Landfried, Christine, 'The European Regulation of Biotechnology by Polycratic Governance' in Christian Joerges and Ellen Vos (eds), *EU Committees: Social Regulation, Law and Politics* (Hart Publishing, 1999).

Lasok, Paul and Rebecca Hayes, 'In the Matter of DEFRA's Proposals for Managing the Coexistence of GM, Conventional and Organic Crops' (2006).

Lee, Maria, 'The Changing Aims of Environmental Liability' (2002) 11 *Environmental Law and Management* 185.

Lee, Maria, 'Regulatory Solutions for GMOs in Europe: The Problem of Liability' (2003) 13 *Journal of Environmental Law and Practice* 311.

Lee, Maria, *EU Environmental Law: Challenges, Change and Decision-Making* (Hart Publishing, 2005).

Lee, Maria and Robert Burrell, 'Liability for the Escape of GM Seeds: Pursuing the "Victim"?' (2002) 65 *Modern Law Review* 517.

Lee, Peter, 'Patents, Paradigm Shifts, and Progress in Biomedical Science' (2004) 114 *The Yale Law Journal* 659.

Leroux, Eric H., 'Eleven Years of GATS Case Law: What Have We Learned?' (2007) 10 *Journal of International Economic Law* 749.

Les Levidow and Karin Boschert, 'Coexistence or Contradiction: GM Crops versus Alternative Agricultures in Europe' (2008) *Geoforum* 174.

Levidow Les and Susan Carr, 'GM Crops on Trial: Technological Development as a Real-World Experiment' (2007) 39 *Futures* 408.

Levidow, Les and Clair Marris, 'Science and Governance in Europe: Lessons from the Case of Agricultural Biotechnology' (2001) 28 *Science and Public Policy* 345.

Levidow, Les, Joseph Murphy and Susan Carr, 'Recasting "Substantial Equivalence": Transatlantic Governance of GM Food' (2007) 32 *Science, Technology and Human Values* 26.

Lewinsohn-Zamir, Daphna, 'Consumer Preferences, Citizen Preferences and the Provision of Public Goods' (1998) 108 *Yale Law Journal* 377.

Lezaun, Javier, 'Creating a New Object of Government: Making Genetically Modified Organisms Traceable' (2006) 36 *Social Studies of Science* 499.

Lieberman, Sarah and Tim Gray, 'The So-called "Moratorium" on the Licensing of New Genetically Modified (GM) Products by the European Union 1998–2004: A Study in Ambiguity' (2006) 15 *Environmental Politics* 592.

Livermore, Michael A., 'Authority and Legitimacy in Global Governance: Deliberation, Institutional Differentiation and the Codex Alimentarius' (2006) 81 *New York University Law Review* 766.

Llewellyn, Margaret, 'The Patentability of Biotechnological Material: Continuing Contradiction and Confusion' (2000) 22 *European Intellectual Property Review* 191.

Llewellyn, Margaret and Michael Adcock, *European Plant Intellectual Property* (Hart Publishing, 2006).

Losey, John E., Linda S. Rayor and Maureen E. Carter, 'Transgenic Pollen Harms Monarch Larvae' (1999) 399 *Nature* 214.

Machlup, Fritz and Edith Penrose, 'The Patent Controversy in the Nineteenth Century' (1950) 10 *The Journal of Economic History* 1.

Mackenzie, Ruth et al., *An Explanatory Guide to the Cartagena Protocol on Biosafety* (IUCN Environmental Policy and Law Paper No. 46, 2003).

Majone, Giandomenico, 'The Credibility Crisis of Community Regulation' (2000) 38 *Journal of Common Market Studies* 273.

Mandel, Gregory N., 'History Lessons for a General Theory of Law and Technology' (2007) 8 *Minnesota Journal of Law, Science and Technology* 551.

Marwell, Jeremy C., 'Trade and Morality: the WTO Public Morals Exception after Gambling' (2006) 81 *New York University Law Review* 802.

McKibben, Bill, *Enough: Genetic Engineering and the End of Human Nature* (Bloomsbury, 2003).

McKibben, Bill, *The End of Nature: Humanity, Climate Change and the Natural World* (Bloomsbury, 2003).

Medina Ortega Report, Temporary Committee of Inquiry into BSE, *Report on Alleged Contraventions or Maladministration in the Implementation of Community Law in Relation to BSE* (European Parliament, 1997).

Meziani, Gundula and Hugh Warwick, *Seeds of Doubt* (Soil Association, 2002).

Micheletti, Michelle, Andreas Follesdal and Dietlind Stolle (eds), *Politics, Products and Markets: Exploring Political Consumerism Past and Present* (Transaction Publishers, 2004).

Murphy, Dale D., 'The Tuna–Dolphin Wars' (2006) 40 *Journal of World Trade* 597.

Murphy, Joseph and Les Levidow, *Governing the Transatlantic Conflict over Agricultural Biotechnology: Contending Coalitions, Trade Liberalisation and Standard Setting* (Routledge, 2006).

Murphy, Joseph, Les Levidow and Susan Carr, 'Regulatory Standards for Environmental Risks: Understanding the US–EU Conflict over GM Crops' (2006) 36 *Social Studies of Science* 133.

Neumann, Jan and Elisabeth Turk, 'Necessity Revisited: Proportionality in World Trade Organization Law after *Korea–Beef, EC–Asbestos and EC–Sardines*' (2003) 37 *Journal of World Trade* 199.

Nottingham, Stephen, *Eat Your Genes: How Genetically Modified Food is Entering Our Diet* (Zed Books, 1998).

Nuffield Council on Bioethics, *Genetically Modified Crops: the Ethical and Social Issues* (Nuffield Council on Bioethics, 1999).

Nuffield Council on Bioethics, *The Ethics of Patenting DNA: A Discussion Paper* (Nuffield Council on Bioethics, 2002).

Oczek, Jeremy P., 'In the Aftermath of the "Terminator" Technology Controversy: Intellectual Property Protections for Genetically Engineered Seeds and the Right to Save and Replant Seed' (2000) *41 Boston College Law Review* 627.

Parry, Bronwyn, *Trading the Genome: Investigating the Commodifcation of Bio-Information* (Columbia University Press, 2004).

Peel, Jacqueline, 'A GMO by Any Other Name . . . Might be an SPS Risk!: Implications of Expanding the Scope of the WTO Sanitary and Phytosanitary Measures Agreement' (2007) 17 *European Journal of International Law* 1009.

Perez, Oren, *Ecological Sensitivity and Global Legal Pluralism: Rethinking the Trade and Environment Conflict* (Hart Publishing, 2004).

Phillips Report, *Inquiry into the Emergence and Identification of Bovine Spongiform Encephalopathy (BSE) and Variant Creutzfeldt-Jakob Disease (vCJD) and the Action Taken in Response to it up to 20 March 1996* (House of Commons, 2000).

Phillipson, Martin, 'Giving Away the Farm? The Rights and Obligations of Biotechnology Multinationals: Canadian Developments' (2005) 16 *King's College Law Journal* 362.

Poli, Sara, 'The Overhaul of the European Legislation on GMOs, Genetically Modified Food and Feed: Mission Accomplished. What Now?' (2004) 11 *Maastricht Journal of European and Comparative Law* 13.

Poli, Sara, 'The EU Risk Management of Genetically Modified Organisms and the Commission's Defence Strategy in the Biotech Dispute: Are they Inconsistent?' in Francesco Francioni and Tullio Scovazzi (eds), *Biotechnology and International Law* (Hart Publishing, 2006).

Poli, Sara, 'Restrictions on the Cultivation of Genetically Modified Organisms: Issues of EC Law' in Han Somsen (ed.), *The Regulatory Challenge of Biotechnology: Human Genetics, Food and Patents* (Edward Elgar, 2007).

Pratt, Sean, 'Schmeiser Renews Monsanto Battle' *Western Producer* 31 May 2007.

Price, Monroe E., 'The Newness of New Technologies' (2001) *Cardozo Law Review* 1886.

Richardson, Benjamin and Stepan Woods (eds.), *Environmental Law for Sustainability* (Hart Publishing, 2006).

Rodgers, Christopher, 'Liability for the Release of GMOs into the Environment: Exploring the Boundaries of Nuisance' (2002) 62 *Cambridge Law Journal* 371.

Rose, Carol M., 'Environmental Faust Succumbs to Temptations of Economic Mephistopheles, or Value by Any Other Name is Preference' (1989) 87 *Michigan Law Review* 631.

Royal Society and Royal Academy of Engineering, *Nanoscience and Nanotechnologies: Opportunities and Uncertainties* (2004, Royal Society).

Sabel, Charles F. and Jonathan Zeitlin, 'Learning from Difference: The New Architecture of Experimentalist Governance in the European Union', *European Governance Papers* C-97-02.

Sagoff, Mark, *The Economy of the Earth: Philosophy, Law and the Environment* (Cambridge University Press, 1988).

Sagoff, Mark, 'Biotechnology and Agriculture: The Common Wisdom and its Critics' (2001) 9 *Indiana Journal of Global Legal Studies* 13.

Salleh, Anna, 'Wearing out our Genes? The Case of Transgenic Cotton' in Richard Hindmarsh, Geoffrey Lawrence and Janet Norton (eds), *Altered Genes; Reconstructing Nature: The Debate* (Allen and Unwin, 1998).

Santamauro, Jon, 'Reducing the Rhetoric: Reconsidering the Relationship of the TRIPS Agreement, CBD and Proposed New Patent Disclosure Requirements Relating to Genetic Resources and Traditional Knowledge' (2007) 29 *European Intellectual Property Review* 91.

Schenkelaars Biotechnology Consultancy, with Risk and Policy Analysts Ltd, *Means to Improve the Consistency and Efficiency of the Legislative Framework in the Field of Biotechnology Article 31 of Directive 2001/18/EC* (Report for the Commission, 2004).

Schertenleib, Denis, 'The Patentability and Protection of Living Organisms in the European Union' (2004) 26 *European Intellectual Property Review* 203.

Schimmelpfennig, David and John King, 'Ag Biotech Patents on the Move' (2005), at www.ers.usda.gov/AmberWaves/June05/Findings/AgBiotech Patents.htm.

Scott, Joanne, 'On Kith and Kine: Trade and Environment in the EU and WTO' in Joseph H.H. Weiler (ed.), *The EU, WTO and NAFTA: Towards a Common Law of International Trade* (Oxford University Press, 2000).

Scott, Joanne, 'European Regulation of GMOs and the WTO' (2003) 9 *Columbia Journal of European Law* 213.

Scott, Joanne, 'European Regulation of GMOs: Thinking about Judicial Review in the WTO' (2004) 57 *Current Legal Problems* 117.

Scott, Joanne, 'International Trade and Environmental Governance: Relating Rules (and Standards) in the EU and the WTO' (2004) 15 *European Journal of International Law* 307.

Scott, Joanne, *The WTO Agreement on Sanitary and Phytosanitary Measures: A Commentary* (Oxford University Press, 2007).

Scott, Joanne and Jane Holder, 'Law and New Environmental Governance in the European Union' in Gráinne de Búrca and Joanne Scott (eds), *Law and New Governance in the EU and the US* (Hart Publishing, 2006).

Scott, Joanne and Ellen Vos, 'The Juridification of Uncertainty: Observations on the Ambivalence of the Precautionary Principle within the EU and the WTO' in Christian Joerges and Renaud Dehousse (eds), *Good Governance in Europe's Integrated Market* (Oxford University Press, 2002).

Secretariat to the Council for TRIPS, *The Relationship between the TRIPS Agreement and the Convention on Biological Diversity* IP/C/W/368/Rev.1 (2006).

Secretariat to the Council for TRIPS, *Review of the Provisions of Article 27.3(b)* IP/C/W/369/Rev.1 (2006).

Secretariat to the Council for TRIPS, *The Protection of Traditional Knowledge and Folklore* IP/C/W/370/Rev1 (2006).
Sell, Susan, *Private Power, Public Law: The Globalization of Intellectual Property Rights* (Cambridge University Press, 2003).
Sell, Susan, 'Intellectual Property and Public Policy in Historical Perspective: Contestation and Settlement' (2004) 38 *Loyola of Los Angeles Law Review* 267.
Sen, Amartya, *Poverty and Famines: An Essay on Entitlement and Deprivation* (Oxford University Press, 1981).
Sevenster, Hanna, 'The Environmental Guarantee after Amsterdam: Does the Emperor Have New Clothes?' (2000) 1 *Yearbook of European Environmental Law* 281.
Seyfang, Gill, 'Cultivating Carrots and Community: Local Organic Food and Sustainable Consumption' (2007) 16 *Environmental Values* 105.
Shaffer, Gregory, 'The World Trade Organization under Challenge: Democracy and the Law and Politics of the WTO's Treatment of Trade and Environment Matters' (2001) 25 *Harvard Environmental Law Review* 1.
Shaffer, Gregory, 'Recognizing Public Goods in WTO Dispute Settlement: Who Participates? Who Decides?' (2004) 7 *Journal of International Economic Law* 459.
Shaffer, Gregory, *When Cooperation Fails: the Law and Politics of Genetically Modified Foods* (Oxford University Press, 2008, forthcoming).
Shaffer, Gregory and Mark Pollack, 'Regulating Between National Fears and Global Disciplines: Agricultural Biotechnology in the EU' (2004) *Jean Monnet Working Paper* 10/04.
Sheingate, Adam D., 'Promotion versus Precaution: the Evolution of Biotechnology Policy in the United States' (2006) 36 *British Journal of Political Science* 243.
Sherman, Brad, 'Regulating Access and Use of Genetic Resources: Intellectual Property Law and Biodiscovery' (2003) 25 *European Intellectual Property Review* 300.
Shiva, Vandana, *Protect or Plunder? Understanding Intellectual Property Rights* (Zed Books, 2001).
Slovic, Paul, *The Perception of Risk* (Earthscan, 2000).
Somsen, Han (ed.), *The Regulatory Challenge of Biotechnology: Human Genetics, Food and Patents* (Edward Elgar, 2007).
Sunstein, Cass R., *Risk and Reason*: *Safety, Law and the Environment* (Cambridge University Press, 2002),
Sunstein, Cass R., 'Beyond the Precautionary Principle' (2003) 151 *University of Pennsylvania Law Review* 1003.
Sykes, Alan O., 'Domestic Regulation, Sovereignty and Scientific Evidence

Requirements: A Pessimistic View' (2004) 3 *Chicago Journal of International Law* 353.

Tait, Joyce and Joanna Chataway, 'The Governance of Corporations, Technological Change, and Risk: Examining Industrial Perspectives on the Development of Genetically Modified Crops' (2007) 25 *Environment and Planning C: Government and Policy* 21.

Thomas, David and Georgina A. Richards, 'The Importance of the Morality Exception Under the European Patent Convention: The Oncomouse Case Continues' (2004) 26 *European Intellectual Property Review* 97.

Ullrich, Hanns, 'Traditional Knowledge, Biodiversity, Benefit-Sharing and the Patent System: Romantics v. Economics?' in Francesco Francioni and Tullio Scovazzi (eds), *Biotechnology and International Law* (Hart Publishing, 2006).

Understanding Risk Team (T. Horlick-Jones et al.), *A Deliberative Future? An Independent Evaluation of the* GM Nation? *Public Debate about the Possible Commercialisation of Transgenic Crops in Britain, 2003*, Understanding Risk Working Paper 04-02.

United Nations Conference on Environment and Development, *Rio Declaration on Environment and Development* (United Nations, 1992).

Van Overwalle, Geertrui, 'Reshaping Bio-patents: Measures to Restore Trust in the Patent System' in Han Somsen (ed.), *The Regulatory Challenge of Biotechnology: Human Genetics, Food and Patents* (Edward Elgar, 2007).

Victor, David G., 'The Sanitary and Phytosanitary Agreement of the World Trade Organization: An Assessment after Five Years' (1999–2000) 32 *New York University Journal of International Law and Politics* 865.

von Bogdandy, Armin, 'Law and Politics in the WTO: Strategies to Cope with a Deficient Relationship' (2001) 5 *Max Planck Yearbook of United Nations Law* 609.

Wallace, Helen and William Wallace (eds), *Policy-Making in the European Union* (Oxford University Press, 2000).

Warren-Jones, Amanda, 'Identifying European Moral Consensus: Why are the Patent Courts Reticent to Accept Empirical Evidence in Resolving Biotechnological Cases?' (2006) 28 *European Intellectual Property Review* 26.

Watson, James and Francis Crick, 'A Structure for Deoxyribose Nucleic Acid' (1953) 171 *Nature* 737.

Weiler, Joseph H.H. (ed.), *The EU, WTO and NAFTA: Towards a Common Law of International Trade* (Oxford University Press, 2000).

Wells, Angus, 'Patenting New Life Forms: An Ecological Perspective' (1994) 16 *European Intellectual Property Review* 111.

Winickoff, David et al., 'Adjudicating the GM Food Wars: Science, Risk and Democracy in World Trade Law' (2005) 30 *Yale Journal of International Law* 81.

Wynne, Brian, 'Uncertainty and Environmental Learning: Reconceiving Science and Policy in the Preventive Paradigm' (1992) 2 *Global Environmental Change* 111.

Wynne, Brian, 'Creating Public Alienation: Expert Cultures of Risk and Ethics on GMOs' (2001) 10 *Science as Culture* 445.

Young, Alasdair R., *Trading Up or Trading Blows? US Politics and Transatlantic Trade in Genetically Modified Food* (2001) EUI Working Papers, RSC 2001/30.

Young, Margaret A., 'The WTO's Use of Relevant Rules of International Law: An Analysis of the *Biotech* Case' (2007) 56 *International and Comparative Law Quarterly* 907.

Zerbe, Noah, 'Feeding the Famine? American Food Aid and the GMO Debate in Southern Africa' (2004) 29 *Food Policy* 593.

Zika, Eleni et al., *Consequences, Opportunities and Challenges of Modern Biotechnology for Europe* (JRC, European Commission, 2007).

Index